CONTENTS

FOREWORD

PART ONE: Brain, Totality and bits & pieces
Concept
Stimuli, Signifiers and Meanings
Theoretical framework

PART TWO: Cognitive activities and Subjectivity
Introduction
Memory
Forgetfulness
Non-conscious memory
Evocation
Concept
Evocative process
Conclusion

PART THREE: Language and the Unconscious Mind
The Unconscious Mind
Concept
The unconscious mind, memory and non-conscious memories
The unconscious mind and subjective gradients
Language
Language, Memory and non-conscious memory
Linguistic stimuli
Language and Evocation
Linguistic convergence and divergence

FOREWORD

As the recently deceased neurobiologist and author Oliver Sacks (1933-2015) used to say in his own words, empirical science, as we know it, does not take the soul into account and disregards the constituent and determining parts of the personal self.

This was to me an inspiring and fundamental idea, which turned out to be very useful, because it prompted me to draft a theory based on mental function, related to the understanding of the root cause and mechanisms by which consciousness –starting from qualitative mechanisms inside the brain– produces and gives supports to meanings. In other words: How does each individual perceive meanings in a unique and unrepeated way in a distinct manner from the rest of the crowd? The aim of this fifty-odd year old theory from the onset of its conception, was to address and rethink (among other approaches) this intricate problem at both conceptual and methodological levels, i.e. by bringing into the study parameters inherent to the semantic nature of the stimulus. These parameters are brought in as a phase of information of abstract characteristics, going beyond all mere physical and chemical aspects, without excluding these from the overall picture.

Since the 20th century the understanding of the brain and its functioning has been growing exponentially, primarily due to the progress and development of new brain imaging techniques. These developments allowed the LIVE visualization of the underlying cerebral activity in both cognitive and emotional life. They also shed light on the puzzling neurochemical nature within complex activities, such as learning and memory. But not on processes related to the genesis of meanings. A singular phenomenon, which for the time being transcends current empirical possibilities. What actually happens is, that the brain –besides operating in neurobiological and cognitive ways, supported by physical and chemical processes–, also imparts meaning to everything, which is perceived or emitted by its own activity. Further clarification is therefore necessary in order to gain deeper knowledge about our essence as humans. My own view is that at the present time we know a lot more about cerebral locations thanks to the remarkable development of brain imaging techniques using state-of-the-art technologies, but in fact we know very little about the consciousness and its role as producer of meanings.

During the second half of the twentieth century or perhaps even before, neuroscience –which at the time was plainly called Neurobiology, or Neuropsychology at the most– used to be a more integrating discipline, that is, more inclined to a interdisciplinary approach. As a result, extraordinary contributions to the study of consciousness at that time were already looming.

The brain is the responsible organ for subjective consciousness, a task that it shares with the semantic nature of all processed meanings. It is a feature of the problem that goes beyond the mere biological domain and still remaining organic. In other words it is the brain, which senses and processes the meaning of stimuli by means of neuronal networks, autonomously from the external surroundings. This is also applicable in the case of attention and perception, which are exclusively regarded as conscious functions.

So, besides the methodological aspect, I also want to highlight both the cognitive as well as the conceptual approach. In this regard a big question arises: Which are the methodological rules that should be applied to do some research into subjectivity, in order to understand its root cause and the resulting cerebral mechanisms? In other words, for the study aiming at a better knowledge of the qualitative signifying processes of subjectivity, which are unique and unrepeatable for each human being, it would have been more fruitful if cognitive contributions around the world and in our country would had been taken into account during the second half of the twentieth century and even before. But unfortunately these were discontinued or inappropriately transferred from one generation to the next.

One of the conclusions I can draw hereby is, that signifying or assigning meaning to identical stimuli, simultaneously by different individuals in diverse and unrepeatable ways, would not involve substantial biochemical changes at neuronal level, as it happens in learning and memory processes. They would indeed exist but they would represent a rather objective phenomenon, common to every brain, and there would be no variations whatsoever for each different meaning in the presence of the same stimulus. In other words, if we signify differently and uniquely from other people in presence of the same stimulus would not cause chemical alterations at neuronal level even if the meanings were varied. Moreover, the core of the problem would not only lie in the intensity thresholds or exposure times to the stimuli or in different protein content syntheses. I am not saying that the mentioned functions are inexistent or are not there for a reason. I would rather emphasize on the fact that different meanings are not necessarily supported by different protein contents. As the research in the field has shown out of observations, however without having explained the causes, subjective meanings are often the result of subliminal stimuli, whose intensity or exposure times are below the necessary threshold to produce cerebral functioning compatible with consciousness, independently from conscious attentiveness and perception. In short, subjectivity would not depend on thresholds, nor on specific proteins, nor cognitive conscious activities, but rather on the unconscious. On semantic rather than exclusively physical factors. At subjective level the subliminal stimuli would contain a higher signifying power and in many cases they would be the actual accountable elements for meaning, which will settle down into consciousness, despite the fact that the latter, –in the role of neurobiological activity– depends on stimuli with sufficient threshold, as shown.

This book also aims at establishing the causes –to be demonstrated– by which the same neuronal circuit is capable of supporting diverse meanings, even antagonistic ones or without relationship with one another. In other words, we are talking about closely associative mechanisms.

As I shall point out in the respective chapter, the difference between memory and subjectivity would lie in the fact that the former is the product of stimuli with threshold and neurocognitive processes, which are common to every normal individual. On the latter however, it is not, since the subjective meanings are instead the product only in those parts of the stimulus prioritized by each particular individual, who allocates a new value order to each stimulus. As we can see, it is different from memory, where the objective meaning of everything will prevail above the meaning of individual constituent parts. In the subjective realm the opposite happens: Not depending on attentiveness, focus and conscious perception.

Hence, the cerebral fundaments of subjectivity would depend on parameters, which are not currently under research. To be more precise, they would depend on the significant nature of the stimulus, one of its physical phases that are not yet being contemplated at empirical level at this stage. In other words, the potential objective or subjective meaning of perceived or thought information, especially of minimal subliminal parts, whose meanings would have semantic inclusion in intimate pre-existing and unconscious meanings, which are in turn in progress of being generated as conscious. Every stimulus is a carrier of objective meaning as a whole. It is an element of parts capable of signifying differently from totality due to a different subjective value allocation, which each individual assigns to perceptions, feelings and conscious thought.

Regarding the phenomenon as such, subliminal parts would have the function to impart meaning at subjective level, as observed through research, but without having proven the actual causes.

The book *"Cerebral Nature of the Subjective Signifier"* is not intended as a tool for declaring absolute truths, but is instead a simple theoretical essay, neither based on plentiful bibliography nor experiments of any kind. As I pointed out before, its main purpose consists of laying the necessary conceptual and methodological groundwork, in order to clarify the underlying cerebral processes involved in human subjectivity. If the reader prefers, it is about a change of paradigm emphasizing more on the signifying nature of the stimuli generating biological and cognitive activity, rather than on their physical properties, also focusing more on the cerebral function they unleash. In conclusion, perceptions, memories and thoughts encompass the totality as a carrier of objective meaning composed of minuscule subjective signifiers. In a sense, this is an autobiographical book of self-reference, since my own life experiences have been analyzed according to this theory and narrated in the book in a manner of introspection and self-knowledge.

Simultaneously with this publication I am releasing the first issue, whose goal is to comment on books by authors of academic excellence and well-deserved international acclaim for their empirical contributions and achievements. I will comment on *The Consciousness in the Brain,* published by "Siglo XXI Editores" Publishing Co., 2014. And I will continue doing so with other authors if time allows.

This subject is intrinsically complex, and I will quote another brief introductory phrase by Oliver Sacks, who as neurologist and narrator had put so much focus on subjectivity: "If a man has lost a leg or an eye, he knows he has lost a leg or an eye; but if he has lost a self—himself— he cannot know it, because he is no longer there to know it."

Dr. MARIO A. TONNINI

April 2018

PART ONE: Brain, Totality and bits & pieces

CONCEPT

In order to understand the concepts of totality and of the individual parts, it becomes of utmost importance to introduce a fundamental independent variable into the research study of subjective consciousness, namely the genesis of meanings by the human brain. As we know, the twentieth century has been prolific in all sorts of discoveries, in particular as far as the brain is concerned. Nevertheless, such advancements and developments of extraordinary scientific rigour, turned out to be insufficient to shed light onto the functional pattern related to subjectivity, whose meanings are unique and unrepeatable on each individual. This involves redefining such problem at theoretical level and so are its empirical methods as well.

Subjectivity was for centuries an exclusive legacy of philosophy and the humanities in general. Neuroscience joined this circle setting itself a goal to unveil its great enigmas but has not yet succeeded in this regard. Subjectivity can –in general terms, –be many things at the same time. It is an abstract summary of everything being perceived and thought, whereby each individual will prioritize –giving up any conscious intent– by questioning himself which parts of everything he or she should choose to make them meaningful. The totality is nothing more than a perception or a conscious idea with an objective meaning. In other words: Which parts of any stimulus –at unconscious level– would be useful to consciousness, so that the latter can generate accessible meanings?

To this aim we first have to understand the genesis of subjective meanings, which are supported by our consciousness: and it is precisely the signifying capacity of stimuli (different on each individual) that will be taken as conceptual basis here, as well as each constituent part of totality, and their ability to assign meanings in diverse ways. This is the reason why the independent variable of this phenomenon would be rather given by the semantic nature of discerned or thought information, instead of its physical and chemical features such as time thresholds or stimulus intensity, ions or specific proteins for each meaning in response to the same stimulus.

As has been noted in research experiments (without having proven its causes), subliminal stimuli are capable of displacing eventual threshold meanings. The brain uses these temporary meanings to produce functional states compatible with conscience, in order to impose its own functional state. As can be seen, stimuli do not make up a whole indivisible totality, since all their constituent parts even without threshold are carriers of subjective meanings of unrepeatable nature on each individual. In a nutshell, subjectivity is a two-edged phenomenon: on the one hand, in neurobiological terms, there are stimuli (with threshold), which trigger a cerebral state compatible with consciousness. And on the other, there are meanings that will not always match those signals. This fact is a paradox because only tiny parts, even infinitesimal ones without threshold, impose their meanings into consciousness.

Evidently we are dealing here with a phenomenon, impervious to the traditional scientific method, whose techniques have proven to be useful for research and neurocognitive objectives, normal and pathological mechanisms, such as memory and learning, just to name a few. Not so much however in the case of subjective contents of human consciousness with autonomous thresholds and chemical changes at synaptic level. This needs to be well understood; it is about neurochemical changes for each meaning specifically. We are under the assumption that the same proteins would support different meanings in response to the same stimulus faced by different individuals. However, we are not only dealing here with a biological difficulty, but with a cognitive one as well, because both conscious attentiveness and conscious perception would not have functional relevance in the signifying action, at least within cerebral signifying activity. Thus, what would vary would be the signifying power of the stimulus as a whole and as component parts.

For a deeper insight into all phenomena at methodological level, there are two essential variables to account for: the phenomenon-independent variable, whose variations will induce observable changes into the investigated fact; in this case the underlying cerebral activity for the referred signifying action. Then comes the dependent variable into our problem raising the question: how does the brain manage signify when it is not relying on thresholds, chemical changes and both conscious attentiveness and conscious perception? In other words, which cerebral mechanisms would have to be triggered to support a meaning lying outside the one of the inducing stimulus in a conscious state of mind? This situation was irrefutably shown in neurobiological and cognitive terms. The missing link here is which part of the stimulus caused the meaning, and by which neuronal mechanisms and through which pathways did this happen? Each brain cell would have diverse gradients, according to the quantity of meanings, by which the corresponding area of the detecting stimulus becomes functional. In order to shed light onto the enigma of subjective consciousness, we should emphasize on the signifying nature of each and every infinitesimal part of totality (the stimuli), as I will describe in successive chapters.

Another dilemma in the study of subjective consciousness lies in the fact that while the brain is operating as an indivisible entity in network mode, its stimuli work autonomously and in parallel, which is totally and partially at the same time. This is the reason why every day we gain more knowledge about the brain, although whatever we do not know, keeps on adding up day by day. This is not the case as far as subjective consciousness is concerned; its conundrums also keep on increasing and we do not know much about them. Naturally, science's tasks involve giving enlightenment to the objective aspects of phenomena, but although human subjectivity is also an objective fact, its contents differ among individuals. In this respect my own view is that the way would be paved by putting suitable emphasis on the semantic causes of subjective experience. This explains why the book starts with "Totality and bits & pieces". It is about totality with a threshold as a carrier of an overall meaning, which is generally objective and can be shared among different people; and bits (parts), which are autonomous from totality and are subjective signifiers. The brain is the most complex organ in nature, and yet the signifying power of each infinitesimal part of the stimuli can sometimes cause its function to lag behind. In a nutshell, the constituent parts of totality, such as stimuli and perception for instance, replicate in all other parts in any of their degrees of fragmentation and even at infinitesimal levels. These parts are essential for the understanding of the cerebral fundaments of subjectivity, on the grounds that they will mean –as a whole or out of context– something different on each individual, time or circumstance.

At this point, I would like to make a digression regarding the unfortunate and paradigmatic case of the patient H.M., who has been quoted in several books. When this patient underwent surgery and in the process had both hippocampi (left and right) removed, he had lost his anterograde memory, and thus the ability to build new memories, –suggesting the importance of the hippocampus in memory and learning– no major developments were made to understand the function of the latter as a cerebral structure responsible for the subjective meanings of unconscious roots. Namely those, which were produced based on infinitesimal parts of verbal and non-verbal information, such as syllables, letters or parts of images (sub-images) as unconscious symbols of highly signifying power. These processes would also be taken over by the hippocampus, perhaps on the basis information out of which memories had been released. These memories could however not linger because of the lack of consolidation, but as information they would still remain active at unconscious level, in order to signify something else or for their eventual recalling (evocation) in future.

Simulation of brain functioning was attempted on the knowledge base of IT (Information Technology), Informatics and AI (artificial intelligence), but they did not deliver the expected results to explain subjectivity. But what really looks promising is the fact that external disciplines, which have little in common with brain knowledge, –although rigorously speaking, nothing is outside brain functioning– such as Economics, Arts and Politics, are already incorporating their particular topics toward a better understanding of neuroscience. For this reason it's important that the connection between Neuroscience and Psychoanalysis also gets established. Fortunately, attempts in this regard are already under way.

Neuroscience as I mentioned before and as we currently know, started in the 1950s and was motivated by the enigmas related to subjectivity, which at the time seemed to be more unfathomable as they actually were. Some phenomena, which today are regarded as highly enigmatic, were in the process of being proven. Art and Literature with their revelations were decades ahead of scientific demonstration, and perhaps this was the reason that prompted and motivated many contemporary scientists to find plausibility in these phenomena. There is the case in one of the novels by the Argentinian writer Jorge L. Borges portraying the memory of Ireneo Funes, one of its fictional characters, or the one from "The garden whose trails fork off" in quantum physics. Or how Einstein by means of his theories made predictions almost a hundred years before their empirical verification. The same applies to published papers by Stephen Hawking, in which he made supporting mathematical contributions to the Big Bang theory. The bottom line is that experimentation is pointless without a consistent theory.

The main goal of this book is to rethink the phenomenon of subjective consciousness in terms of its root cause and the underlying brain activity. It is about a change of paradigm in the study of the nature and genesis of conscious subjective meanings, which will enable us to formulate a theory of mental functioning. As I pointed out before, subjectivity is above all, a paradoxical activity. It is a singular phenomenon with opposing edges. On the one hand we have the brain as a biological substrate common to every normal individual, and each individual signifying or assigning meanings differently, on the other. This means in other words, behaving uniquely and without repetition in response to identical stimuli under normal conditions.

It is a great paradox and the most likely answer would be that every stimulus carrying an objective meaning encompasses an infinite amount of parts, most of which are subliminal and could mean different things. And those parts displaying different levels of fragmentation, even reaching infinitesimal sizes, replicate in all other stimuli. Therefore, functional parts to the most diverse variety of meanings will intervene according to the rank and distribution imparted by each individual to everything perceptual or cognitive. It works in a similar way as memory, but in this activity unlike in memory, the objective meaning of totality will prevail, as compared to single parts. This will obviously happen at conscious level, because at unconscious level brain cells detect each part replicating it in other stimuli with other meanings, some of them memorized, others subjective. In short, we can say that every individual facing a stimulus would reorder and classify the bits and pieces of totality according to his/her subjective preferences or priorities. That is why not all information gets encoded as memory, due to the fact that each individual would only select, –regardless if consciously or unconsciously– the thought or perceived parts of the stimulus which he would need to retain in memory. In other words, it is about assigning meanings according to his needs and expectations. At subjective level it is perhaps worth noting that identical parts responding to the same stimulus will vary their signifying character in each individual.

Hence, the subjective would take place on the basis of an unconscious selection, out of parts of totality based on individual priorities or needs. Thus, every individual would only take what is necessary to signify or to supplement a pre-existing meaning (e.g. in the process of conscious gestation), without damaging its threshold or other physical and chemical properties of the stimulus. Allegedly, since in some cases the meanings would derive from totality, while on others only from some of its fragments. What is real is that the stimuli, like everything else, – however different– replicate into other stimuli by means of their constituent parts just by varying their distribution and signifying power. In other words, while every stimulus as a whole contains an objective meaning, which can be shared amongst various individuals, its fragments have the particular feature of signifying differently in each case, circumstance or timeframe. Hence, the subjectivity content would only depend on how each individual handles the reshuffling, i.e. sorting of the stimuli's or sub-stimuli's fragments. From this follows that the meanings, even foreign ones but consistent to fragments of totality, are associative among themselves, because – as we shall see further on– parts which are common to various stimuli would act as a bridge among different concepts, including opposites. As we can see, the same neuronal circuit is capable of supporting the most diverse meanings, even if they display affinity among themselves. But, what are the links joining them? We could assume, as I mentioned before, that common parts of the same or different stimuli would do the connecting work, even having signified differently. For example, a vertical line could either represent the capital letter "I" or a flag pole, i.e. the stimulus remains the same, but its meaning has changed. In a nutshell, the same information, which in the first place created subjective opposing meanings, ended up associating them as information.

This is how subjective experience comes into play. On the one hand we have signifiers continually mutating and in the middle of change processes, and on the other a brain function, which is common to every relatively stable individual, acting as a substrate carrying dissimilar meanings depending on the case. This will not necessarily involve the synthesis of a different protein for each meaning. Summarizing, the variable of subjective diversity would lie in the signifying nature of the information received, as opposed to the brain activity generated at biomolecular level. The fragments or parts of totality that has been perceived, thought or recalled would then behave as vectors of subjective experience. And every fragment will then become a new integrator of parts, which after assigning meanings would get recycled and become available at conscious level, in order to carry on signifying the same or different things.

Totality represents the supplied information "fitted" with adequate thresholds in order to trigger brain activity. Nevertheless, if its meaning lacks of subjective importance, each infinitesimal subliminal part would take over the functional command to impose conscious meaning. For a better understanding, let's think it this way: a subliminal stimulus would impose a meaning provided it has the minimum required subjective importance. For this reason intensity, frequency or minimum timing required for a stimulus to trigger brain activity are inadequate variables for showing subjective consciousness. This however doesn't mean that the stimulus can do without a threshold, as I will illustrate further on. When assigning meanings the stimulus' physical threshold would be inversely proportional to its signifying power.

Stimuli comprise infinite amounts of parts without thresholds of their own. Furthermore, we have a problem when trying to explain how –as already observed– subjective signifiers can be subliminal stimuli, lacking the necessary potential for being detected and thus trigger consciousness-compatible brain activity. Here we have to understand that the term "subliminal" is a concept that actually has more than one connotation. And two in particular are worth mentioning: First, when the physical intensity does not reach the lowest limit (limen) to be able to generate brain functioning, and second, even when reaching the minimum required limit, it lacks signifying possibilities, because it will be subliminizing more information in the process. Possibly within the cerebral substrate of subjectivity, some mechanisms could exist that would allow to revert or invert thresholds, or else could be temporarily "lent to" other subliminal stimuli, in order to replace a particular meaning. In this regard, I will formulate some hypotheses further on in the book.

Finally, within the brain's signifying action we find two types of thresholds, which are inversely proportional to each other: one is related to its physical intensity and the other is related to its signifying value. On the latter threshold, the higher its signifying value, the lower the intensity it would require to assign meanings. As we can see, this is a very complex phenomenon, which has the potential to be addressed at experimental level. On the other hand, if we want to prove mechanisms within conscious memory, stimuli with enough thresholds are needed as a minimum basic requirement for a percept to be memorized. Memory, otherwise cannot take place. It is obvious that in response to subliminal stimuli, neither perception nor conscious memory will be present at all. The subjective does not rely on the stimulus' thresholds. And it neither depends on more focus directed to the perceived situation.

Unlike memory, which relies on conscious stimuli, subjectivity has unconscious roots and this is what makes it more difficult to understand its mechanisms from the standpoint of functional patterns, as is the case with neurocognitive objective mechanisms, namely the assumption is incorrect, in that we are filtering out all irrelevant parts of the information, in order to focus into what is essential for the meaning to be memorized. Subjectivity does not follow conventional rules, which apply to the objective functional brain. Furthermore, subjectivity is paradoxical and person-specific. The fact that certain parts of the information are at subliminal level does not mean that the brain would discard them, but it will filter them out of the conscious plane instead. This is because they are not of fundamental importance for memory, but they are still valid to signify at subjective level. It is true that we are not utterly conscious in response to everything what is thought or perceived, however, the unconscious state of mind will be in charge of receiving and preserving all the parts it will use (or reuse) as meaningful units. And last but not least, it will prevent mind and consciousness from being interrupted during the waking state as well as during the sleep or dream phases (regardless if awake or asleep).

Signifying action at subjective level does not depend on the reciprocal laws of excitation, but rather on states of matter or phase transitions, at which the responses are inversely proportional to stimulus size. Subjectivity is paradoxical, since it should presumably work according to the mentioned transitional states of matter, at which physical thresholds are not taken into account in certain states of matter. In physics we understand states of phase or phase transitions as changes in the state of matter, according to the state in which they happen to be, or the ones they adopt. In this case, the stimulus phase is represented by the signifying phase.

As can be seen, the meanings of consciousness would only depend on semantic values of the selected information in each particular case, even if these were subliminal. This is how hidden colours or syllables within a greater volume of information generate meanings, which displace the meaning of totality from consciousness as well from the rest of the fragments, regardless how objective the meaning actually is. A radically proven fact is that the unconscious perceives, memorizes and becomes subjective based on another kind of stimuli, at which thresholds do not count, and when they articulate amongst each other they will have to guarantee the temporary continuity of both mind and consciousness. No matter how irrelevant stimuli are or seem to be, the truth is that our brain never discards any of them. As a result, in identical percepts there will be an infinite amount of signifiers, which after assigning meaning will eventually recycle as stimuli to keep on signifying something else. Obviously, the objective meaning of what is perceived will remain unchanged, but it will not always take place within consciousness. Subjective experience can be memorized, even though its stimuli might not reach any conscious entity or absolute threshold. Beyond brain cells or neurons, absolute thresholds also abound plentifully in the brain. These thresholds –as we will see further on– sustain meanings without much subjective importance or at least meanings in the process of conscious gestation. That's the reason why a syllable, a letter, a dash or a single colour can signify without the need of a threshold of its own, and may be hidden within a unit with another meaning with threshold.

Because the formidable literary work of the Argentinean writer Jorge L. Borges cannot be omitted in this instance, I would like to make a new comment about the author: if we all were like "Futnes, the memorious" (one of his fictional characters) we would no longer be in the position to think because memorizing everything at conscious level would be cognitively incapacitating. This would let us assume that the brain would be in possession of those minimal parts, so that it will no longer have to perceive or memorize them to signify complementarily or diverse ways. I also learned in my readings that Borges had a prodigious memory, which did not disable him to think or compose. He perceived every detail like nobody else. These fine details were in the end the seedlings of his literary creation. "Jorge L. Borges used to transform everything into literature" as the Peruvian writer Mario Vargas Llosa recalls. In July 2016 Pablo Gianera, Argentinian essayist, critic and columnist wrote an article entitled "Borges, the reader of himself". I can quote a passage which reads: ".. . Borges used to read letter by letter as if he would be spelling the words. One of the examples is "Emma Zunz".. .", he concludes.

In each chapter of the book I will address various cognitive activities such as memory, evocative processes and language, among others, from a subjective viewpoint. Another author worth quoting in this context is Umberto Eco (1931-2016), literary critic, philosopher, semiotician and university professor. In his opinion he said referring to the world wide web that "The problem will be focusing on the hierarchyzation of information, which will emerge out of the web's depths. In other words, this huge ocean of signifiers will always have to be maintained and kept in a logical order to make sense". And then I ask: Who else will do the job other than the brain?

In a nutshell, totality and constituent parts would have to signify in different ways; THAT at objective level, and THESE, at subjective level, even being opposites. Totality tends to join objectivity and subjectivity. A piece of furniture, for instance, is perceived as a sort of objective totality, which can be included in many objective and subjective contexts at conscious level as well. The unconscious would instead work by means of infinitesimal parts of that totality as subliminal parts of the furniture piece or of the context it belongs to.

Subjective meanings depend to a greater extent on the semantic nature of the stimulus than its intensity thresholds or time of exposure. This happens in such a way that despite being subliminal stimuli, they can also signify at conscious level. Another important point to be considered is, that there are no specialized areas in the brain for each meaning, but rather regions, which get enabled in an all-embracing network, while all meanings begin to take shape. But what do exist are engrams to identify meanings, such as happens in memory. Engrams are means by which memories are stored as biophysical or biochemical changes in the brain in response to external stimuli. The cerebral signifying or memorizing action is non-specific but a rather complex function, which enables the encoding of remembrances and concepts. Subjectivity is an exclusively human evolutionary sign and is related to diversity, whose nature is not only biological but semantic as well.

Each area of the brain is an assembly of neurons that can be enabled anytime in response to stimuli, which are compatible with their respective functioning. For example, let us take the fusiform gyrus, which is brain region responsible for recognition of faces, for instance. It gets enabled at an early age in response to the presence of familiar faces. Hence, we already have our first dilemma: a face is an image carrying one or several meanings, but also infinite amounts of other things not related to faces at all. A face is an assembly or unity composed of an infinite amount of conscious and unconscious parts. These constituent parts have a capacity to signify differently from the unity, allowing us to infer that besides the fusiform gyrus activation, it will also impact on the functioning of other brain areas, which are responsible for other meanings. The brain –as all integrated units– works in a series of networks processing signifying stimuli, whose meanings compete against one another, so that they acquire conscious state. Parallel to the unity, each part of the stimulus will also signify in a parallel way, but diversely from the unity.

One of the brain characteristics, which should be taken into account, is its capacity to store and process vast amounts of information. We are talking about levels of ca. 10^{14} bits, or an equivalent of 114 *Terabytes.* It is an eloquent figure if we think about the huge amount of information to be stored. A substantially higher capacity than any electronic system that has so far been designed. What is striking is the fact that the information has an exponential growth rate, which is due to two reasons: First, the information brought in from external sources through perception, and second, its intrinsic capacity given off from its own neuronal activity, produced on the basis of infinitesimal parts of totality. What's more, it is about constituent parts, which after having signified something remain available to resume signifying, but not necessarily the same content. In other words, the encoded information that we understand as memory, including its constituent parts, do not lose their signifying power once they completed the signifying process. And they are not attached to or locked to a specific meaning either. This implies that the brain is continually appending information, perhaps in a redundant way in terms of stimuli but not in terms of meanings. This is an evolutionary sign of the human brain, which cannot be reproduced in robotics.

In a nutshell, it is worth remembering the experiences of Wilder Penfield, an American-Canadian neurosurgeon, who noticed on his patients that when specific neurons were stimulated, they were able to recall situations of their lives, or memories, which in many cases they could not remember having memorized or experienced. This is a singular phenomenon, a fact from which we can infer two things: the unconscious is as active or even more active than consciousness, and what's more, it does not forget. On the contrary, it can take over forgetfulness at conscious level. Stimuli are entities, which carry objective meanings. And these can be memorized. Furthermore, they encompass signifying parts, whose "memories" (mostly subjective) would generally remain in the unconscious, waiting for the right conditions to arise in order to gain conscious state; or they just break in directly into consciousness displacing meanings, which have responded to stimuli as a whole. The most curious aspect of this is, that meanings of two different activities can be related to each other within the same neuronal circuit, even without semantic or temporary affinity. Instead, they are linkable by means of minute common parts (replicated) in each meaning-producing stimulus. This stimulus can be either objective or subjective, as we will see further on in the next chapters.

In this instance it is appropriate to do some fiction, which in most cases is superseded by reality. What would happen if all possible parts of signifying information would associate among themselves? The answer is: quite many things. What is not going to happen, however (if this was your expectation) is that the brain would only work at conscious level without the unconscious. Exactly the opposite would occur: the unconscious would eventually expand becoming the new absolute totality, which can be fragmented into new parts. But please, I am not trying to frighten the reader, as this is only fiction. What is non-fiction however is that without any solution of continuity, every new part will erect itself as a new totality. The bottom line is that without the unconscious, consciousness, as we know it, would not be possible.

Mario Bunge, an internationally famous Argentinian epistemologist analyzed for decades the big controversy between psychoanalysis and neuroscience. I personally suppose that both disciplines have some common point of convergence, as pointed out by Michael Gazzaniga, the founder of modern neuroscience. Furthermore, neuroscience and psychoanalysis will not reveal the complex unknowns inherent to human subjectivity by going separate ways. These unknowns are impossible to understand in holistic terms if know little about its cerebral foundations and semantic implications.

STIMULI, SIGNIFIERS AND MEANINGS

In terms of cerebral activity, information can be interpreted as stimuli. On the other hand, when we talk about signifiers we have to understand which one of the selected parts of the stimulus will generate a meaning for each individual according to his/her subjective priorities. When we address meanings, we have to find out the resulting concepts out of each and every part, among which only one of them would eventually attain conscious state. Now raising this issue –because everything we perceive or think represents an objective totality, which will not necessarily be remembered–, most parts of that totality will be stored as non-conscious memories at conscious level. This singular activity is essential for signifying at subjective level, as we will see in later chapters. Every stimulus, as I pointed out before, is a carrier of objective meanings comprising an infinite amount of parts, which are neither dependent nor anchored to a particular meaning. They are instead configurative of a new totality and are also functional to other meanings. This particularity explains why each individual will signify differently in response to the same stimulus according to his/her subjective priorities and needs of adaptation.

Every part of totality, regardless how small, minute or infinitesimal, will build a new totality, whose meaning would be capable of displacing the meaning of an original objective stimulus from consciousness, and replacing it with a subjective meaning. The functioning of the brain during the signifying action depends on countless factors. Especially on its semantic nature, not only its physical characteristics, but also the intensity threshold or time of exposure to a stimulus. As a result, the totality and constituent parts are of fundamental importance in our consciousness of the world, whose goal would be to allow the individual to get rid of direct stimuli replacing them with abstract ones. In other words, the meanings of subjective consciousness would only depend on abstract stimuli, which derive from parts of concrete stimuli. Thereby the abstraction would be rather a resulting function of subliminal signifiers in terms of information.

Ironically, most parts of the stimulus being used by the brain to signify are relatively lacking of thresholds and other functional variables for conscious activities. As we can see, information furnished with intensity thresholds and with sufficient action times do not seem to succeed in imposing the meanings supported by subjective consciousness due to their subjective irrelevance. It is an activity largely induced by biological, physical and chemical factors. Thus the information would be replaced by the meaning of some of its parts, even if these lack the necessary threshold. This is in fact a phenomenon, which has a catch, since both the conscious as well as the unconscious planes work in parallel. What is astounding and paradox is the fact that a stimulus lacking of subliminal threshold is incapable of generating consciousness at biological level, but only to impose the meaning relativizing it in such a way to interfere with attentiveness and perception.

This for instance can happen to anybody standing in front of a text or image and perceiving only words, even if these are subliminal or if they underlie subliminally from totality. Our unconscious however, had to see them in order to complete an idea or concept, even though they have little or nothing in common with the objective meaning of that text or image as a whole. Therefore, those parts –"captured" by subjective concepts under development– would have made easier its conscious conversion.

It is a clearly substantiated fact that thresholds of subjectivity would be inversely proportional to their intensity and time of exposure to a stimulus. This is perhaps due to the fact that their semantic nature constitutes a different state or phase of the stimulus capable of (paradoxically) challenging the laws of biology, which are inherent to the reciprocity of excitation. Consequently, the lower the intensity of the stimulus, the higher the relevance its semantic features would acquire, if it awakens subjective interest in a particular individual. In a nutshell, it's about subjective signifiers, which while getting articulated amongst them, would lead to the continuity of the mind and its contents, namely conscious and unconscious meanings. A singular function which would take place during both the waking state as well as deep sleep, and during dreams or so-called oniric phases.

In conclusion, information has two functions: first to promote the necessary biological mechanisms for consciousness to emerge out of cerebral functioning and second, to sustain its meanings, which is not a minor task. Totality contains an infinity of signifiers of something else. Each part becomes a new totality. That is the reason why the meaning assigned to the same stimulus would differ in each circumstance and timeframe, not only among different individuals, but on the same individual as well. Let's take for example the blue colour: regardless of which image is perceived or recalled, or the word "blue" used to designate the colour by verbal means, will signify by itself without modifying the objective meaning of the image, word or original thought. But it would eventually displace it from consciousness due to its lack of significance, so that a subjective experience can take its place. As we can see, a single colour or letter can transport us to other meanings. Consider for instance a masterpiece or the colour we would choose to paint our home, or the necessary word to complete a line of text for a poem. Consequently, the function of the signifying elements would be to pull back cognitive events of the past into consciousness. These past cognitive events could be remnants of memorized or non-memorized experiences stored in the unconscious. Hence, the more insignificant a stimulus, the higher would be its signifying spectrum due to its association to an infinite amount of unconscious meanings. The brain does not discard information, if it does not use it immediately, it will do so in future activities. It could also use it as an evocative signal for apparently forgotten memories.

If we talk in neurocognitive terms, the memory is the largest and richest source of signifiers with a capacity to generate autonomous subjective states of all conscious remembrances. In this manner memory and subjectivity would have an analogous, though not identical functioning. The differences lie in the fact that stimuli of a subjective experience do not require a threshold of their own. On the other hand, conscious memory is not possible if the precept lacks a threshold. Subjectivity would be due to the signifying action of infinitesimal parts of information, i.e. in some of its fragmentary grades. This is how a little line could represent the eyebrows of a face or the edge of a table, among so many other meanings. Thus, identical information is functional to all precepts or ideas, so its meaning will vary.

A new example comes in handy, so that we can withdraw from so much theory: a simple table is a concrete object, and as a stimulus equipped with the necessary threshold, it will allow to be perceived, memorized and evoked at conscious level. But its meaning will vary in accordance with the context in which it would be perceived or memorized. As a result, once it is out of the context of totality it would have at least two functions, first the evocative, in order to recall gatherings with family or friends (objective facts) and second, to signify on its own at subjective level. Another possibility arises if we feel too tired and need some rest, then the rectangular shape of the same table (perceived or thought) would rather remind us of a bed, or when appropriate, we could use it to lay down. Or if we need to get across a river stream, it would be very handy if we could improvise a bridge or a raft out of it. What happens is that parts of the information, which were caused by memory are out of context and can have another meaning, no matter if they have a threshold or not, because they could use the one from memory, or the one from the subjective needs under conscious development.

As a result, each perception or thought has overall meanings, which include an infinite amount of non-storable components at conscious level as signifiers of something diverse. These components in the brain will become subcomponents of information, which we may call sub-precepts and sub-cognitions, and their function will be to link the stimulus as a whole (the actual trigger of conscious access) with the subjective meaning (to be supported by consciousness). Any form of linguistic expression (verbal, spoken, thought....) such as words, syllables or letters may induce different meanings from the original sentence or idea, regardless if they have semantic affinity or not. One syllable of any word, for example "NO", or the letters "N" and "O" separately could prompt to reduce the "tone of voice .. because it's disturbing..." or ".. .DO NOT do sOmething that is NOT prOper". Subjectivity is paradoxical, as it is not governed by any logic whatsoever. We never cease signifying, let alone during our sleep. It is of fundamental importance to have the power to think up or to get a concept independently from objective facts, even for better memory consolidation. This is due to the simple reason that the necessary stimuli to get a lasting memory do not require any physical thresholds, but signifying power instead.

THEORETICAL FRAMEWORK

The study of subjectivity dates back to ancient times and was the heritage of several disciplines such as philosophy, psychology and humanities. This was the case until recent times, during which Neuroscience, owing to the astonishing progress in brain knowledge, set itself a goal at unveiling its functional role within subjective life, although with mixed prognoses. While some of the highest qualified authors claim that the phenomenon could only be unveiled in about fifty years, there are others –not so optimi*stic*– *who* argue that it will never be revealed. As we can see, there seems to be no consensus of opinions; on the one hand we have the relatively *optimistic ones,* and the ones who are supporters of the mysteries behind the phenomenon on the other. And the standpoint of the latter is plagued by pessimism. Its advocates are among others, well-known psychologists and philosophers, like the philosopher Colin McGinn and the cognitive psychologist Steve Pinker, whose central argument is that the mind cannot understand itself. Karl Deisseroth, an American psychiatrist and bioengineer, who is also a well-known researcher of the biomolecular foundations of memory, declared: "It may be the case that we might never succeed in getting to understand the brain". He certainly understood that neurochemical mechanisms would not be sufficient to explain the operating principle of the subjective mind. Therefore I suppose that unlike the optimistic ones, the pessimistic authors overestimate this conundrum. I personally think that it is possible to unveil it, if and only if we change the neuroscientific research paradigm, so that we can reveal the problematic in a shorter timeframe as expected.

Steve Pinker put forward an argument declaring that the human brain, as well as the remaining species is a product of evolution and its cognitive clarification would be limited. This is partially true. Of course, "Dogs –as he put it–, cannot understand maths". (By the way, most humans do not do either, and here I include myself). What is true as an argument is that dogs because they lack the capacity of abstraction –though I am not so certain about it–, could never understand a mathematical equation, but this disability does not mean the impossibility for the human species to understand subjectivity from the perspective of brain functioning. In my opinion, it is precisely by means of abstraction that we could start examining its neural twists and turns, as the illustrious philosophers of antiquity did with the unfathomable enigmas of the human mind, such as the importance and the active intervention of consciousness.

But in this dilemma we should rather be cautious and not get into extremes. We should seek balance and carefully formulate our opinions. The brain is a product of evolution, but this does not discard the possibility that someday, –which could be near– we could explain the mental operating principle. Since last century, times of discovery have tightened drastically and as a result, knowledge has ever since grown exponentially.

Despite such opinions related to the great difficulties involved in unveiling subjective consciousness, there is no doubt that without a brain, both mind and consciousness would never be possible. Even with all difficulties and empirical limitations involved to clarify the unfathomable enigmas of the human mind by means of traditional scientific methods, I think there are other ways to crack the nut, for example, by using theory and inference. Since the second half of the twentieth century and even before, objective knowledge of the brain as a regulating system of the body and as a biological site of complex cognitive activities grew at an incredible pace. But certainly it did not so to enhance the knowledge on the brain's capability to signify differently, in an unrepeated manner on each individual. And this is essential to understand the human mind.

The brain works based on stimuli, which are either perceived or put together within its intricate neuronal circuits. Perhaps the largest obstacle to clarify subjectivity is not to discover the underlying neurochemical mechanisms, –which after all are common to every normal individual, as we will see in the following chapters– but in understanding the nature and essence of the stimuli triggering unique and unrepeated meanings on each individual. The true nature of the stimulus that could explain subjectivity is of semantic character, and is not tied up to the conventional laws of physics, chemistry and biology, but rather on other rules that, while still remaining physical, follow other functional patterns. What's more, the neurochemical mechanisms sustaining subjective meanings would not substantially differ from those (already discovered or to be discovered) underlying in various cognitive activities. Within the cerebral action of signifying, diversity is due to the variety of meanings that every stimulus can hold in itself and even within the same stimulus after a change of times or circumstances. Subjectivity means semantic and conceptual diversity. Therefore, multiple meanings in response to the same stimulus would have (as biological justification) the same chemical activity. Irrespective of the above, there is still a great deal to be discovered in this field, as is the case with short or long term memory, of meanings or episodes supported by identical ions and proteins. Subjectivity does not require any stimulus thresholds, as its mechanisms are unconscious. The unconscious works with other kinds of thresholds and stimuli.

In a nutshell, the underlying cerebral function in subjective consciousness in response to the same stimulus would be common to every normal individual. This is perhaps the only objective pattern in the subjectivity problem. Although memory varies from one individual to another in terms of meanings and episodes (including the same objective remembrance varies for each individual), it does not happen in the same way at ionic or proteic level, because in this instance they are common to any memory. Thus, diverse subjective meanings would also be supported by an analogous functional framework. In conclusion, we can say that proteic synthesis is fundamental for brain function, but the actual variable for subjectivity is semantic, independent from thresholds and from chemical changes specific for every meaning. In the end what we should find out is the reason why identical signals signify differently on each individual, while the neurochemical function for every meaning remains unchanged.

Another aspect should be taken into account and this one would prove that the current methods applied by conventional neuroscience to unveil underlying mechanisms in every complex cognition are not suitable for this purpose: every individual signifies differently from others in response to the same stimulus and at the same given time, because every individual would have selected a different part of the stimulus, even subliminal without participation of attentiveness or conscious perception. But the one of non-conscious memory is indeed appropriate as a source of signifiers, which must not necessarily be linked to the conscious meaning of remembrance.

The stimuli of subjectivity would thus have two sources, namely the outer world and the memory. The brain's storage capacity and information processing is immeasurable and displays exponential growth. Every part of totality is functional to unconscious meanings in parallel to conscious ones. The brain detects the totality of perceived or thought information as everything and as signifying components, which after signifying will remain active to signify something else. Thus, finite information can generate an infinite amount of meanings. The brain does not filter out information, since the one, which has not been used will be saved in a kind of unconscious repository for future signifying purposes.

As we can see, subjective consciousness and cognitive activities in general, with memory amongst them, are closely related. Memory is the outcome of all information encoded in the brain as a synthesis of everything that has been perceived, including the subjectivity of every single unconscious part thereof. Memory is so to speak, the tip of the iceberg, while its non-visible part, as irrelevant as it may be to be memorized, is the part that would eventually generate subjective experiences. Parts replication will increase as stimulus fragmentation increases. The replication of those parts will occur in higher quantities than the information, which has been thought or perceived. For example, in order to devise something, to bring forward a concept or communicating, identical linguistic signs, phonemes and syllables will be used. But in a different order they will convey a new word with another meaning. The same will apply to shapes and colours, since by varying its distribution, a new image with another meaning will result. This is one of the reasons why a neuronal circuit can support more than one meaning, even opposite ones. Therefore, mind and consciousness would only have temporal everlastingness without any gaps.

PART TWO: Cognitive Activities and Subjectivity

. **Introduction**

. **Memory**

. **Forgetfulness**

. **Non-conscious memory**

. **Evocation**

. **Concept**

. **The evocative Process**

. **Conclusion**

PART TWO: Cognitive Activities and Subjectivity

INTRODUCTION

Subjectivity and memory are, in terms of brain function, closely related activities. However, their stimuli, mechanisms and conceptual contents differ substantially. Memory depends on stimuli with threshold, while a subjective meaning by contrast, can be the consequence of subliminal stimuli. In order to be able to memorize, each individual will make a summary of whatever he or she has perceived in order to retain its objective meaning. One of the most common ways to subjectify is to carry out the process from memory, that is, parts of totality of the stimulus that generated the meaning. Here we are dealing with an important difference between objective remembrances and subjective meanings. Nevertheless, subjectivity –not being a memory in the strict sense of the word– can also be memorized and recalled because it depends on those subliminal components of totality. Those subliminal parts in a way would simulate memory's mechanisms at unconscious level. In a nutshell, memory and subjectivity are the consequence of information or response to stimuli, but within memory stimuli will have to be perceived at conscious level. Subjective meanings instead would depend on another functional plane of the brain, namely the unconscious. At cerebral functioning level everything is relative, in that it is also possible to memorize non-conscious events. In other words, subliminal stimuli or without conscious threshold could be memorized at unconscious level. Brain information can neither be lost nor forgotten, but its meanings by contrast, can both be lost and forgotten!

In short, memory is generated on the basis of a synthesis of objective information, while a subjective meaning will largely depend on the decoding of the stimulus (broken down in its constituent parts). These are implicit sub-stimuli built into the inducting information of remembrances and its meanings, perhaps with scarce functional prominence within memory. When percepts are detected, new stimulus clusters are being formed in the brain, and these clusters would eventually link the overall stimulus with the resulting subjective meaning. In other words, new stimuli called sub-percepts or sub-cognitions (depending on their origin, from the outside world or when they are part of perception) detach themselves from cognitive activities, from ideas or from autonomously generated thoughts out of attentiveness and percep- tion. While memory is a conscious process being forged on the basis of threshold-bearing information, the subjective depends on subliminal information fitted with signifying power. All concepts and subjective experiences can also be memorized like objective percepts, but unlike those, at unconscious level instead.

Any percept, object or image –such as a house–, contains a meaning, which can be memorized objectively by countless numbers of people. What is striking is the fact that each part of the percept "house" replicates throughout all possible information; and even if it lacks meaning, sense or threshold, it will be able to signify autonomously from totality. Let's look at another example, such as a window frame. As a subliminal stimulus at conscious level hidden from the rest of the parts of the house, it could induce a subjective autonomous experience of its overall meaning, such as the colour of a garment. The curious thing is that both meanings are parallel; on the one hand we have the house (objective), and the garment (subjective) on the other, so that the underlying neurochemical activity would be the same, so that it would not experience any substantial changes whatsoever. This peculiar feature allows to put forward the following hypothesis: once a subjective meaning, capable of being memorized, attains the conscious state, it could also –if reinforced– spawn proteic synthesis in the mid and short terms. In the following chapters we are going to emphasize on this particular topic of utmost importance, as well as on other far-reaching aspects.

As we can see, the differences between memorized objective and subjective meanings, built upon minimal parts of totality triggering remembrances, are due to the signifying capability of the stimulus rather than to the generated neurobiological activity. And this is precisely what objective neuroscience strenuously seeks to show, namely in trying to identify the proteins (if any), which are common to any brain, and which would not show the causes of non-repeatability of meanings. In other words, it's about identical proteins at neurochemical level sustaining unrepeated meanings on each individual even in response to the same stimulus. But actually it would no longer be the same, but rather different and autonomous parts disguised in information, and even the same or identical part will signify differently according to the case. In the same way as several and diverse pleasure-inducing stimuli activate identical brain areas –for instance love, addictions or cellular phones–, also meanings would vary accordingly, triggered both by identical stimuli and bio molecular activity.

As a result, neurocognitive activities in general, such as memory and learning are governed by information as a whole and as constituent parts. Not everything we perceive is stored in memory but it does get stored in the brain. We should remember that parts of every percept in a fragmentary degree are replicated systematically in all possible information whether perceptual or cognitive. So we have two differences: Firstly these get diversely distributed in each totality, being therefore functional to specific objective meanings; and secondly, when they free themselves out of that formally organized structure, they end up signifying something else.

As I pointed out previously, information should not be confused with meaning. Although every stimulus can draw our attention, its meaning will not always necessarily be in accordance with our own subjective needs or interests. This is precisely the particularity I would like to emphasize in order to understand the difference between a conscious remembrance and a subjective experience. In order to signify at subjective level we would only take some parts of the event, which can be (as I pointed out before) regrouped into sub-percepts or sub-cognitions. These would act as an inter-cerebral link – a sort of cognitive limbo– between a perception, a recalled memory or idea and the resulting subjective meaning. As I said before, these are functional links that would also take place within memory but with the difference that those would be more conscious and the intent would be measured.

This is why our brain not only holds memory but non-conscious memory as well– an activity, which will eventually signify differently from conventional memory–. This lies beyond the fact that forgetfulness would not be as such, and that evocation would not only have the function to retrieve remembrances, but also to convert unconscious subjective meanings into conscious meanings. The information reaching our brain is detected simultaneously and in parallel by subjective meanings in the process of being generated at conscious level. These meanings, once captured, would increase their gradient of brain activity, so that they can attain the conscious state. The flow of stimuli inside the brain is relentless and the unconscious is in a permanent state of expansion, mutation and change. These processes play a major role in subjectivity.

Works of true excellence have shown immense knowledge about memory. And although there are many unknowns which need to be unravelled, on a day by day active brain areas during memorizing processes (neuro-imaging) are being unveiled by means of imaging techniques, or their supporting neurochemical processes are being identified. In a nutshell, while memory depends on the processing and overall synthesis of the percept, subjectivity depends on the individual's own analysis and his own unconscious selection of totality. In other words, the parts of non-conscious memory contained within in every percept, memory or cognition do their own thing with the conscious remembrances. Parts that make up all images or the signs and linguistic symbols, which are implicit in all ideas, whether verbal or not, build up non-conscious memories of outstanding signifying power. In other words, sub-images contained inside any possible images or words, syllables or letters become inducing symbols for ideas. Consciousness is a temporary continual activity in terms of information, but with discontinued meanings undergoing a mutational or changing process, on the grounds that every infinitesimal part of totality is a more complex totality, than the original with the same components, but distributed in a different way.

Each part of coded information as memory, will be recycled to signify something else (of what we are unaware of) every time it has signified. Our brain is not a machine, which has been built to detect and memorize, but maybe to store and process information, in order to signify from its infinitesimal constituent parts. Non-conscious memory, as we shall see in chapters ahead, is any element or part of perceptual or thought totality, which in its most diverse fragmentary degrees will signify differently from objective totality, even if it has not generated a conscious remembrance. What is then the reason for such a tremendous importance assigned to the cognitive life of non-conscious memory? Among other reasons, it is because all external information perceived or thinking based on cerebral (internal) information, each part is detected simultaneously by neurons, whose function is to process that same underlying information in other meanings.

MEMORY

Memory is a cognitive activity whose objective mechanisms are being proven at empirical level without solution of continuity. It represents a neurocognitive activity of utmost importance, in which classifications and definitions are commonplace. Although they are synoptic and schematic, they offer the advantage to allow their sub-categorization into diverse classes and types. Some theorists point out that memory consists of three phases: sensory or acquisitive phase, consolidation phase and evocative phase. Nevertheless, remembrances are not only a product of what has been perceived, but also of what has been thought. What is paradoxical is the fact that remembrances would not be easy to consolidate just by repetition as seen in experiments, but also by means of associative or evocative signifiers, even without a threshold. That is why evocation, beyond being an essential stage for the retrieval of memories, it also encompasses a far broader neurocognitive spectrum: we not only recall memories but also subjective elaborations or concepts, which never had memory entity status, but they were memorizable indeed. In fact, memory would rather be part of evocation.

Neurobiological experiments of great excellence such as those carried out by Eric Kandel laid the foundations to categorize memory in terms of its permanence in hours, months, years or life-long durability. But in reality those elapsed times do not correspond with real or concrete life at non-empirical level and are not to be correlated with Freudian psychoanalysis, as pointed out by

Kandel, researcher Emeritus at Columbia University, NY in one of his books entitled "In Search of Memory" (W. W. Norton & Company 2007). Both Freud and Kandel were born in Austria and therefore the latter's interest at a young age in understanding Freudian theory. He pursued this to such an extent, that in the mentioned book he declared with wit and irony, that once he would settle down in the U.S. his first goal would consist of identifying which parts of the brain corresponded anatomically and physically with the ego, the superego and the id. And it was there where he set his permanent residency following a brilliant scientific career to become a Nobel Laureate in 2000. But he could not achieve this particular goal he had in mind. Of course, Freud was referring to subjective facts, while Kandel was pointing to objective mechanisms.

Resuming our classifications about memory, there is one, perhaps the main one, which subdivides them into two distinct groups: declarative and non-declarative memories. The former are related to conscious remembrances of meanings, concepts and life episodes, which are part of our own biography, hence they are also called autobiographical. In fact they are associated with subjectivity, which is the core subject of this book. The latter are the so-called non-declarative memories, also known as implicit or procedural memories, and are regulated by different brain areas than the former.

The main site of memory in the brain is the temporal lobe, whose development at human level allows the brain to store vast and unfathomable amounts of information at both conscious and unconscious level. It is in this process, in which the hippocampus' role would be to sort out the existing chaos in the cerebral cortex, and beyond its proven interference in memory and learning, it also plays an essential role in unconscious stimuli in order to signify at subjective level. The question is how? It would include every subliminal signifier into the totality of possible meanings, among which only one meaning would eventually access consciousness, while the rest would remain in a "waiting list".

As we can see, subjectivity is as complex or even more complex than the brain itself, which says a great deal. There is a higher number of signifiers than neurons, because when we talk about identical information and the activity of a single neuron, the meaning will be different without any variation of the underlying neurochemical processes within each subjective concept. On the one hand we have complex bio molecular changes, which are common to every normal brain, even some activity levels being shared with diverse animal species, and abstract concepts, unrepeated for each individual, exclusively human on the other. So, memory is about two things: information and meanings. It is about information that at the same time is consolidable and unconsolidable, but not in degrading terms, as the name suggests, but rather tending to cognitive enrichment.

Due to this and other reasons, the memory problematic should be addressed again and rethought. Although memory's mechanisms may be objectively common to any normal brain, its meanings, conscious per objectives and shareable among different individuals, are full of subjectivity. Hence, there could be a classification of objective/ subjective, and verbal and non-verbal memories. The essence of meanings from memory is always subjective, regardless how objective meanings actually are. Neurocognitive research should not only be restricted to or subject to show objective activity patterns. This would be like trying to explain the contents and meanings of billions of books, studying their format, paper characteristics, typesetting, typography, etc. Or a chair's features as a piece of furniture made of particular types of wood.

In conclusion, memorizing involves two things: First, to perceive the essential part of a stimulus at conscious level "filtering out what is unnecessary" ; and second, at unconscious level to process those minimal details, which are generally "filtered", but not discarded with the aim to focus on the understanding of their objective meaning, in order to signify differently from memory.

Memory and subjectivity merge in semantics, which is their turning point. The former merges from information as totality, while the latter from each of its constituent parts capable to signify differently according to the case and circumstance. As can be seen, meanings can be memorized or can be reconstructed outside perception, and can also be memorizable from pure cerebral activity without relying on influences of external stimuli. Episodic or autobiographical memory also gets linked to subjectivity due to the individually specific circumstances of our lives, which is unrepeated as meanings and in our fellowmen. In any event, subjective meanings in their highest purity level only take place when they are product of infinitesimal unconscious parts of the stimulus, whether they have been memorized or not.

Summarizing, memory is an essential cognitive activity, and it has probably been the major target of research and discovery from the second half of the twentieth century to the present day. It is an essential activity in the learning process teemed with definitions and classifications that although synoptic, allow it to be subcategorized into different classes. One of those mentioned classifications, which is perhaps the most accepted nowadays, points out that memory consists of three phases: the sensory or acquisition phase, the consolidation phase, allowing the retention over time and the evocation phase, which becomes active every time information and its meanings are retrieved at conscious level. Certainly, nothing is absolute and even less in neuroscience, since we do not only memorize by means of perception but also from our ideas and feelings, emitted from the own autonomous cerebral activity of sensory elements.

Consolidation does not only imply the mere repetition of information as it occurs at empirical level, but from the action of linguistic parts and from images, which although functional to different meanings being consolidated, reinforce it by means of unconscious mechanisms. This happens even less concerning evocation as an exclusive base of memory, because it is also fundamental in all aspects of cognitive life; we do not only recall remembrances but also subjective elaborations and reasoning, among other activities, which never had memory entity. As we can see, every part of the classical phases of memory have connotations in the totality of cognitive and emotional life. If we remain true to reality, memory would be one of the phases of evocation and not the other way around.

As has been pointed out right at the beginning of the current chapter, memory has been the object of neurobiological and cognitive research of great excellence. Some were awarded with the Nobel Prize like the work of Eric Kandel, who obtained this distinction at the beginning of the 21st century for his discovery of neurochemical processes underlying short and long-term memories at Columbia University. But in reality, as declared by Kandel in his book "In search of Memory", the duration of remembrances verified in experiments do not correspond with real life. And neither does the knowledge about the brain making parallelisms with psychoanalytic postulates. I bring up this digression since Freud and Kandel were born in Austria, and while there was an age difference between them, they were contemporaries and therefore Kandel's interest in psychoanalysis at a young age before moving to the United States to pursue a brilliant scientific career.

Besides of the memory classification in terms of temporal permanence, there is another which is well-known dividing it into two categories, namely, declarative or explicit memories and non-declarative or implicit memories, also called procedural. The former are related to conscious remembrances of the memory's meaning in question (semantic). The latter are linked to life episodes (implicit or procedural) and are instead related to neuromuscular abilities including the capability of positioning and orientation in space. Both types of memories are regulated by different cerebral structures, and the declarative memories are the ones with the strongest relationship with subjective life due to their conscious nature. Their brain location is the medial temporal lobe or MTL, in which the hippocampus plays a fundamental role in their production. It is also a cerebral structure, which would also intervene in the capacity of subliminal stimuli to signify from unconscious activity.

In a nutshell, on the one hand we find complex neurochemical mechanisms of objective character, which are common to man and to any living being capable of memorizing, and abstract concepts, only to be memorized at human level, unrepeated and diverse in each individual on the other. This is the turning point at which memory and subjectivity are merged functionally, and this does not trigger any changes in the underlying neurochemical activity at all. This is the reason by which the essence of memorized meanings is subjective, while their underlying cerebral mechanisms are objective in nature.

There is a recurrent statement of modern conventional neuroscience, in charge of unveiling the underlying cerebral mechanisms in cognitive and emotional life, which are common to every normal individual: In order to be able to memorize, we filter out the irrelevant perceived information. This is partly true at conscious level, but it is not the case at unconscious level. This is a state, in which the brain does not filter anything but instead stores everything, in order to signify from each and every part of that totality. Strictly speaking we remember what had been selected at conscious level, beyond the fact that at unconscious level the brain captures subliminal stimuli it would need for the conscious conversion of the meanings, which are still being forged.

Summarizing, meanings can be memorized either from perception or without depending on it. While memory is the product of the synthesis, subjectivity on the contrary, depends on parts of the information out of totality, which had been memorized.

FORGETFULNESS

Subjectivity is a cognitive state closely linked to memory and forgetfulness. In my own understanding the forgetfulness concept should only be confined to the conscious plane, since the unconscious would be in charge of preserving, re-signifying and process the retrieval of the information at different times and circumstances. Alternatively, a memory, which has been forgotten due to the lack of consolidation, –in neuroscience known as a short-term or short duration memory– would have turned to signify something else, such as subjective concepts. The unconscious would be the plane of cognitive activity that among many other functions would be in charge of preserving the forgotten information and its meaning associated to other events. To forget would be a mode in which the brain retrieves the information, in this case involved in a forgotten memory, so that this can be reused to signify something else, or to recall it amidst suitable circumstances.

There are indeed memories, which are more forgettable than others. Unequivocal occasional data would be less retainable than others. There are different types of memories, which should be differentiated from one another, such as the remembrance of a date at which an historical event has taken place, from other facts within that same memory, which are perhaps less forgettable than the date itself. Human memory is not permanent in that dates, telephone numbers or very specific data cannot be remembered accurately, but this does not imply to forget the circumstances associated with them. Let's refer to a well-known example quoted in books and conferences by many authors: forgetting a telephone number after it has been used, as was the usual case in earlier times. This was common before we were able to store it into the memory of a device allowing us to reuse it when needed. The problem arises when the device breaks down or when we lose it. Of course, in the past, numbers were recorded in a diary or index; this was after all, a similar way of handling the information. What happens is that not remembering the number, when we need to make a telephone call (working memory), does neither imply to have forgotten the purpose of its use, nor the recipient of the call, a fact that would be of great concern. Stated briefly, consciousness, memory and forgetfulness are closely related activities. Analogously to false memories, there is also a false forgetfulness! Where some of them are caused by inadequate evocative circumstances.

The human brain does not keep everything in memory. We have to understand this concept properly; we are not talking about memory as the capacity to store information. And to forget does not imply to lose the totality of remembrances, including memories, which did not last over time (so-called short term memories), whose meanings, no matter how ephemeral they were, could have been associated with other re-signified remembrances, or they became consolidation signals of other semantic or episodic remembrances, which are required to last. Another possibility is that the neuronal circuits, which are responsible for the forgotten memory, –identifiable with high precision today– may not be available because they could be busy with another function. Every circuit is common to an infinite amount of objective and subjective remembrances. But something, which has not yet been proven, is why dissimilar and even antagonistic meanings are able to settle down within one same neuronal circuit. One of the reasons could be that meanings without semantic or temporary relationship among themselves do associate by means of common parts of their inductive stimuli, which in one memory they have one meaning and on others mean something opposite or different. It could also be the case that the mentioned part could be in the process of developing a subjective meaning, autonomous from memory that could cause a temporary recall of those memories; and this is what gets unreasonably often assigned to forgetfulness, but in actual fact we are dealing with a false forgetfulness. Therefore, not retaining in consciousness certain aspects of a memory, lies in the fact that the information functionally involved in remembrances could be busy with another process forging a subjective experience, since stimuli can not only be evocative but signifying as well. Under normal conditions, as far as the essential remembrance is concerned, forgetfulness is never absolute. If human memory would be permanent, the unconscious would not make any sense at all, and everything would be just memory.

What is true however, is that over time memory weakens –something inevitable and normal– especially when the remembrance did not get consolidated or when it has lost its emotional commitment. During the recalling process of a remembrance, as well as in the remaining phases of memory, stimuli will remain coded inside the brain, so that they get decoded and recoded at a later stage in functional parts to ideas or to subjective experiences, which over time become less and less like the originally memorized content. Once the memorizable stimuli are coded inside the brain, they will be decoded as signifying stimuli, and this will trigger an exponential growth of the unconscious, which in turn will increasingly host more and more signifiers. Parts of the information would then follow an inverse functional path to the one they had when they were constituent elements of totality with an objective meaning to be memorized. In other words, centrifugal impulses from the centre to the periphery, would relegate the functional participation of attentiveness, perception and memory to a secondary plane.

Forgetfulness goes through multiple disciplines, among them neuroscience and psychoanalysis. While for neuroscience forgetfulness is an objective phenomenon dependent on a more or less complex interconnectivity and neurochemical exchange at neuronal level for its consolidation, temporary durability and for its association with other memories, for the Freudian psychoanalysis it is rather related to defence mechanisms of the ego, such as negation and repression, which are related to forgetfulness, but are not necessarily the same. In order to consider a proven fact in science, it has to be reproduced under identical circumstances, as many times as it may deem necessary. Forgetfulness, however, never follows identical causes and many of them cannot be reproduced at experimental level, therefore we should be careful not to generalize the concept. Memory and forgetfulness represent an indivisible reproducible unit at experimental level, but not in semantic and subjective terms. We should not discredit experiments that unveiled and were able to prove the neurochemical nature of memory, learning and forgetfulness, but we should admit that under natural circumstances in life there are also other intervening parameters, which are impossible to be put to test by means of conventional methods and techniques.

Summarizing, for psychoanalysis the concepts of memory and forgetfulness are a bit opposed to those of objective neuroscience. Memory and forgetfulness are two different faces of the same coin, not only subject to physical and chemical, but also to semantic processes. Brain imaging techniques are without a doubt a fantastic tool designed to show the underlying cerebral activity in memory and forgetfulness, but not to identify their meanings. It is obvious that forgetting involves having memorized something in the first place, but since memory is an activity associated by affinity and through antagonisms with other memories, it is also possible that whatever has been forgotten reappears as if by magic, especially when we recall something else or when we think about something without any relationship with the forgotten concept.

Among the memory classifications, one of the most well-known are the ones already mentioned, which subdivide them into explicit or declarative memories –autobiographical–, and implicit, non-declarative or procedural memories. The former are fundamental to retain in consciousness meanings and life episodes, which we add during our lives. And their infinitesimal parts behave in fact as unconscious subjective signifiers. Our brain is capable of and has a great capacity to store and process a far higher amount of information than any robot or cybernetic equipment, but in qualitative terms it does so by using minimal parts of totality to signify or to recall.

Memorizable stimuli are not just prickings or electric currents experienced by lab animals, who learn to avoid them, thus being rewarded to retain those stimuli within memory. We nevertheless not only memorize what causes us pleasure. Unfortunately whatever is traumatic and painful is as much or even more memorizable, and perhaps less forgettable. Certainly, there are pain-related experiences, which animals memorize and learn to avoid, while we humans under certain circumstances have no choice but to endure. Stimuli are bits of information containing objective meanings, which are signifying parts of something else, whose potential meanings would also be memorizable. These parts repeat themselves as I pointed out before, in all thought or perceived information, and from unconscious activity they contribute to the self-consolidation of diverse memories. Let's consider for example, a straight line or a slanted little line; as information they are common to any abstract or concrete image. But as a potential agent of consolidation or as a meaning it will vary on each individual, in spite of being subliminal or if it were disguised within a larger volume of information.

Brain imaging and other techniques are suitable means by which underlying cerebral mechanisms in memory, learning and forgetfulness can be visualized. However, the meaning of everything learned, memorized or forgotten will –in terms of meanings– vary from one individual to the next, because the suitable variable for its further study and research is not only the neurochemical activity but also the signifying nature of the stimulus. Of course, without proteic synthesis and other neurochemical and biological processes it would not be possible to memorize, learn or forget. Anyway, forgotten memories would be neither absolute nor irreversible, at least from totality; but maybe the original version of the remembrance does, and this also happens with memories that are never forgotten despite having undergone changes. Nothing is unchanging inside the brain, because every neuronal circuit can sustain memories from different times without any semantic relationship among them.

What often happens is, that forgetfulness is usually explained as the extinction of a condition. And it is not the same. In his excellent book entitled "The Art of Forgetting", Ivan Izquierdo an Argentine-Brazilian scientist (born in 1937) and a pioneer in the study of the neurobiology of learning and memory, clearly depicts this phenomenon. But I understand that extinction and forgetting are two different things. Extinction is related to the weakening of a particular function conditioned by the lack of reinforcement, while forgetting is related to meanings. Here we are dealing with two concepts: on the one hand we have the stimulus and the neuronal circuit, and the meaning of a conditioned function, which was wiped out on the other. Stimuli are not restricted to a neuronal group but to the entire network instead. Furthermore, the same extinguished circuit sustains an infinity of meanings or concepts, but only one meaning cannot pull the others together. Memory is in fact many things: information, mechanisms and meanings, which are not necessarily tied up to conditioned processes. The stimulus or information is functional to a great diversity of meanings. Beyond that, every stimulus, as basic as it may be, is an element of an infinity of signifying parts. The same neuronal circuit can sustain diverse and antagonistic meanings and from different times. Explaining forgetfulness as the extinction of a conditioned function, may be connected to inadequate empirical references from real life.

Experimental attempts to induce "therapeutical forgetfulness" by means of chemical and electrical procedures deserve here additional comments, and even narrowing down in trying to delete the most traumatic parts of a remembrance. As I mentioned before, memory is not pure hedonism, so we therefore have to rethink the lab experiments with animals, but focusing instead on inferences to the human kind to whatever has been proven. Memorizing and learning involve fundamental molecular changes – protein syntheses –, and there is no doubt about that. Nevertheless, man can do without concrete concepts in order to generate ideas or to remember, replacing them with abstraction. The attempt of permanently "erasing" remembrances for therapeutical purposes –although technically feasible– would neither be effective nor permanent, or by means of minimal electric currents because identical parts of stimuli in other meanings would facilitate in short intervals to return to what has been recalled and even to recuperate lost or blocked proteins. This approach is experimentally feasible by using protein antagonists for memory consolidation.

This comment comes in handy, since there are publications of this kind in prestigious journals like "Nature" about works on this particular phenomenon. In a nutshell, notwithstanding whether these attend to or not to ethics –which is not the purpose of this book– the forgetfulness achieved would not be permanent under normal conditions. All memories, besides neurochemical processes at synaptic level do carry meanings, which are subject to multiple evocative conditions, in other words biological and semantic processes are not governed by the same laws. Although every traumatic emotional situation is treatable at clinical, neuropsychiatric and psychotherapeutical level, it would not be effective to be erased in the medium and long terms. The unconscious does not forget, since implicit signifiers of any memory would be able to restore it within a brief period of time at the mentioned activity level. Even a single point, a line or a specific colour would have enough evocative power to recall supposedly forgotten facts, due to their extremely broad signifying spectrum.

Memories are not erased without reason. Unlike a software, the brain's information IS indeed erasable. Although we know that it is not easy to erase certain data from the web, brain function cannot be simulated neither by AI nor by any other computational technology. All conditioning is substitutive, and ceasing to reinforce it will not lead to total forgetfulness, but only at conscious level. An infinite amount of unconscious signifiers underlie in every conscious memory and they are capable of modifying their evocative circumstances, so that these can be restored back into consciousness. In conclusion, our brain can rebuild forgotten remembrances under normal conditions.

The Argentinean writer and essayist Jorge Luis Borges somewhere in his vast work refers to forgetfulness without even mentioning it. In Shakespeare's Memory, it reads: ".. .every new writing overwrites the previous one and this in turn gets overwritten by the next one, but the almighty memory can unearth any imprint no matter how temporary it has been, if enough stimulus is given". Further on in his work he added: "I had Shakespeare's memory in a latent manner, i.e. the reading and re-reading of those old volumes that the stimulus was seeking".

NON-CONSCIOUS MEMORY

Non-conscious memory is an activity related to memory and consciousness. It is not the opposite of memory, but instead those parts of memorized information, which after having contributed with the forging of remembrances became part of an unconscious functionality as evocative signals, associative among memories, or as signifiers of something that never formed a memory. For better illustrating this concept, non-conscious memory is composed of those non-conscious parts of perceptions and cognitions, which in spite of their assumed irrelevance in memory, they can mean something diverse when they get out of the context of totality. If we consider that memory is the cerebral encoding of whatever has been perceived, non-conscious memory would be the decoded parts of that totality, integrated at unconscious level. In other words, it is like each of the parts of a puzzle, which has not yet been assembled, and whether assembled or re-assembled it will represent an image, whose meaning will finally be whatever remains in memory. One thing is the stimulus and the other its meaning. What is really interesting is that every part, which went out of context of, the overall perceived or thought information, will contain a diverse evocative, signifying and associative capacity in itself.

Our brain is composed of thousands of millions of neurons, whose main function entails to process all parts of the overall stimulus in all its fragmentary levels. Stimuli related to brain function differentiate among themselves as totality, but are composed of identical parts, although diversely scattered in each possible totality, this means these are replicable in every stimulus and with different functions in each brain cell and the underlying neuronal circuit in one memory. This is the reason why each memory is sustained within a complex integrating neuronal network of infinity of new neuronal algorithms, and it is understandable to suppose that they would function in parallel with conscious remembrances. As we can see, since cerebral information is composed of identical parts of stimuli of different meanings, diversely distributed, it is finite and even redundant. The main difference lies, as a result in that one same part means diversely in every totality, in each individual and even under changing times and circumstances. In short, while information is finite, its possibilities for signifying in diversity tend to infinity. Therefore, the human brain's storage capacity of information in qualitative and quantitative terms is, – although untidy and chaotic– infinitely higher than that of any electronic equipment.

If we go back to the puzzle example, the resulting image after the assembly would represent what we commonly retain in our conscious memory, while every building block of the game will signify something else. For example, if the image corresponds to a landscape in which we have a small house at the shore of a lake surrounded by lush and flowering trees, each and every one of these building blocks, as well as their intrinsic parts (the leaves of a tree), will signify something else as an objective fact, while their colour could mean a possible tone that we may choose to paint a wall or when we choose an item of clothing (subjective facts). The water of the lake could even remind us of our seaside holidays. When we signify, those minute parts attach themselves to one another joining events of different times. A house remains a house, whether trees at the shore of a lake surround it or not. Memory and subjectivity are not the same, but they do merge at semantic level.

Our brain has an enormous capacity to receive, detect, store and process information, of which minimal parts are actually used, namely sub-percepts and sub-cognitions, to signify simultaneously and memorizing a synthesis of totality. This means that parts of what has been perceived, are grouped into sub-percepts or sub-cognitions, and they would have evocative or subjective functions. Thus, there is a link between the objective meaning of a perception or thought and what has been recalled, in case of being a memory, or a meaning as a subjective experience, although only one meaning at a time will enter consciousness, the remaining ones, in parallel would accordingly increase their gradient or cerebral activity level in future. This is the reason why a non-conscious memory would not depend on physical thresholds, but on cerebral gradients used as semantic substrates, which after reaching a certain level would provide the necessary threshold to install a particular meaning into consciousness without any intervention of attentiveness or perception.

Finally we can say that non-conscious memory is signifying information that even being unnoticed, takes part in the memorization of meanings, autobiographical episodes and in the forging of our system of values and beliefs. It is true, as Stanislas Dehaene points out in his book "Consciousness and the Brain", that information without sufficient or subliminal threshold lacks enough power to generate a cerebral state compatible with consciousness, but what is also true is that stimuli fitted with a threshold are not always carriers of subjectively transcendent meanings. This is the reason why the signifying responsibility would only fall upon a minimal part of the stimulus. The mind would be a sequence of signifying stimuli and discontinued meanings. It is common to observe that we can wander from one idea to another one that does not make sense without noticing it at conscious level.

The cerebral function of subjectivity would take place during phase transitions, in which neither stimuli nor responses depend on egalitarian laws of reciprocity of excitation, which generally govern neurobiological functioning. These laws clearly define that, the scale of the response depends on the signal's intensity, i.e. the more intense the signal, the higher the reaction would become and vice-versa. The biological processes are governed by the mentioned laws, but not the signifying action of the brain, which would only depend on the phase or state of the stimulus, in this case the semantic, far beyond physical thresholds and neurochemical changes specific for every meaning. These states had already been studied by Pavlov to monitor wakefulness and sleep by proving that weak stimuli (or subliminal) are paradoxically capable of generating intense responses and vice-versa.

It is therefore thinkable to suppose that if subliminal stimuli cannot signify by themselves at conscious level, they can indeed do so by using alien or "borrowed" thresholds. Perhaps, idle underlying thresholds in meanings without any subjective importance. But here we have the great paradox: while threshold-bearing stimuli are required to generate consciousness, this is not necessarily true for meanings. The cerebral function, which acts as a substrate in the signifying action, as well as the underlying neurochemical activity and the thresholds are not enough to unravel the enigmas of subjective consciousness. So the signifying nature of stimuli is the key.

Human consciousness is derived from mechanisms, which are common to every brain. This does not apply to its meanings, whose biochemical activity would not vary in response to subjective diversity, or in response to the same stimulus. So far, we have no empirical evidence thereof. In view of this problem, consciousness would be a chain of cerebral mechanisms common to every individual, but with different meanings in each case, even in response to the same stimulus. This fact allows us to assume that the same neuronal circuit is capable of sustaining diverse meanings caused by identical stimuli, not as an integrated totality, but for some constituent parts. Subjectivity is value-relevant and semantic at the same time. It is about a singular activity in which everything is not the sum of the parts, but a new more complex totality instead. This is the reason why an individual A will signify something and individual B something else for having selected different parts of non-conscious memory of the same stimulus with another order as totality. Paradoxically they are all different composed by identical parts.

In short, each memory encompasses as non-conscious memory infinite amounts of unconscious signifiers. Concepts of a different meaning are subjectively linkable by common parts or characters, which can be either linguistic or non-linguistic. A single letter immersed within a linguistic context of greater extent is subliminal as information, subliminized or occluded from the rest of the set. The same happens with faces and all kinds of images, even artistic ones. Hence, it is not unreasonable to assume that each component of an abstract painting would correspond with an artist's individual life experiences, independent from the time. In functional terms, in our brain every event is linkable to others out of common parts, because in some fragmentary state every percept or cognitive activity is handled as replicated identical information.

Thus, any subassembly of parts, such as letters randomly grouped without any logical sense or without building words with meaning, such as four consecutive consonants is part of non-conscious memory, which would impart subjective freedom to our consciousness. For example, the linguistic sub-percepts "JCGZ" and "JCZG" are formed based on the flood of verbal and non-verbal information, in which we are submerged. These could help us recall the surname Gómez for instance. Which would have been the determining letters to spark our evocation? Obviously G and Z, but not necessarily in the order as they appear in the stimulus, since the brain can invert it, according to subjective priorities or by "skipping" or changing the sign (C by G, due to the figurative similarity). It would then be plausible to think of Gómez in response to "CZ", a subliminal part of "JCZG" –even if they are central parts not of beginning or end of subpercept– instead of GZ or "JCGZ", although the latter bears a higher threshold.

Jorge Luis Borges confirmed it this way. In his "Library of Babel", he mentions: "I am not able to combine certain characters e dhcmrlchtdj, which had not been foreseen by the divine library, and in some of its secret languages would not hold make a terrible sense... ".

Ultimately, infinitesimal equal or similar parts, which correspond to two or more can generate a subjective meaning from subliminal parts or from parts, which have been subliminized within a greater context of information. Subjectivity would then be consistent to qualitative articulation processes among infinitesimal signifiers of what has been perceived, felt or thought up to that moment without any connection among them. It is an activity far more dependent on non-conscious memory, than on attentiveness, perception and conventional memory (at conscious level). Therefore the unconscious would be, as if that were not enough, capable of memorizing its own language and thresholds. One could possibly ask oneself whether it is about a function subject to analogous processes of conscious memory, with the difference that its mechanisms would rather depend on infinitesimal parts of whatever has been perceived or thought. Everything thought or denied could also be consolidated as any memory at conscious level. Does forgetting in fact involve the irreversible loss of a remembrance, or is it plainly the absence of a suitable evocative signifier? We will see. It has been demonstrated though that a memory varies from one evocation to the next. Perhaps the cause, by which successive evocations re-signify or modify the same remembrance, would be that in that particular memory evocations insert themselves in the passing of time associating unconscious signifiers of something else. Therefore, the diversity of meanings would be directly proportional to the degree of fragmentarity of the selected element of totality in order to generate it, and inversely proportional to its threshold. In other words, the more seemingly insignificant a stimulus may be, the higher its possibilities of subjective unreapeatability.

EVOCATION

Concept

The evocative process fulfils two important functions: to retrieve remembrances and to express subjective meanings related to our own ego and to the surrounding world, with relative autonomy of attentiveness and conscious perception. For this reason, it is essential so that consciousness can sustain meanings and concepts. It is a functional neurocognitive activity aiming at guaranteeing the continuity of consciousness, so that the latter does not experience any interruptions or existential loopholes during the states of wakefulness and sleep. When we recall, our brain mobilizes tremendous amounts of cerebral and cognitive information related to overall remembrances and unique subjective experiences, which are unrepeated on each individual.

The evocative process may have either conscious or unconscious origin. It will only become conscious when the recalled subject is a product of threshold-bearing stimuli; and it will become unconscious when it takes place from minimal parts of the stimulus, generally without sufficient threshold or subliminal parts. It is certainly not a paradoxical phenomenon, in that conscious intentionality not always mediates. Memories do –among other reasons–re-signify constantly, due to our own subjectivity.

The responsible information in our memories at cerebral level is in a continual decoding process into unconscious parts, which are ultimately stimuli triggering subjective meanings. Besides, the more intense the fragmentarity becomes, the more the resulting parts would reach infinitesimal and quantum levels and therefore a higher evocative spectrum. This happens because the parts are replicated in the other stimuli, i.e. a dot or a dash is implicit within every bit of information capable of generating memories or to signify. These symbols would be detected by homologous parts of other meanings, which in turn will try to strive to acquire conscious state.

If we take colours, for instance, elementary shapes or linguistic characters, which besides being abstraction signals in their own right, are relatively foreign stimuli or independent from the words and the source images, which would signify differently and uniquely on each individual and under different time circumstances. As long as the stimulus' fragmentarity increases as a whole, the resulting parts will not become necessarily less complex: and they will not be more subliminal than in the first fragmentary stages. In this sense, each letter getting out of context of a verbal totality would have more signifying possibilities than a group of words with a unique meaning. In a nutshell, every part no matter how subliminal it becomes within a greater context of information, would be received by identical parts, which are replicated in all information with other meanings exerting their influence upon their cerebral gradients in order to gain conscious state. Every part of totality including subliminal fragments would, as a result, displace the conscious meaning produced by the integrating stimulus of the part in question. Hence, a possible way of defining consciousness could be in evocative terms, that is, an activity in continuous process of mutation of signifiers and meanings.

THE EVOCATIVE PROCESS

Because everything is connected with everything, we could explain the evocative function as follows: we could resource to geometry as a formal science, whose goal is to study elementary and complex shapes. A line, for instance, is the shortest distance between two points, a stimulus that may have active influence in the genesis of remembrances and its respective recalling function. In other words, a line as an indefinite or definite succession of points in evocative terms is a multi-signifying stimulus, for example of the horizon or the edge of a table. This means of different meanings and even paradoxical in character. Therefore, identical parts of the same stimulus do match with one another somehow in a fragmentary degree with different meanings, and the more elementary or insignificant as a stimulus it might be, the higher its evocative power. As we can see, minute parts of an objective totality still without sufficient threshold, bear in themselves a very high signifying and recalling power. In conclusion, the stimulus' general functions are the following: to reinforce or consolidate meanings within memory, to recall them or to signify something else, different from remembrance. In any case, the intensity threshold of the signifier as a stimulus would not contribute much to generate a subjective meaning, but on the other hand it would do so, by generating brain functioning compatible with the conscious state, as was proven at neurobiological level.

Going back to elementary geometry, which is a discipline that allows us to exemplify situations of every day life: if we draw or imagine a diagonal between the opposing vertices of a square, then the figure will then become split into two right-angled triangles of equal area, with common hypotenuse and equal length cathetii. Furthermore, it would act as an angle bisector into opposing right angles formed in the square. In case of two diagonals, the resulting triangles will be four, and their respective hypotenuses will then become each side of the original square. Therefore, as we go along and keep on adding new elements to the original shape, new parts will emerge, and this will not cause any modification to the square or its meaning. Nevertheless simple strokes or little lines do modify their objective meanings due to the appearance of new shapes and images. This means that a fragmented totality in constituent parts would have several objective and subjective meanings. What actually happens is that, every geometric shape shows substantial analogies to real life in that common parts to each percept will signify differently from the latter. Images would be at conscious level less forgettable than words, since their constituent parts are relatively captive at unconscious level and their eventual meanings would be merged with totality, and as a result, their autonomous signifying power will be blocked at conscious level. But this does not happen with parts of the language, because syllables and linguistic signs carry their own signifying power and their inclusion into multiple ideas and concepts. Linguistic signs will display a higher divergence power in other meanings in connection with parts of the images, which in a recurrent way would converge into the original image.

Accordingly, everything and component parts represent evocative signals of concrete images and abstract meanings. Stimuli and meanings are not the same. Meanings re-signify at each evocation. At conscious and unconscious level, stimuli represent evocative symbols inducing abstraction of whatever is tangible. A circle allows us to recall various things starting from a "wheel" to the concept of "roller bearing". Or perhaps remembrances in which roundness would have a central importance, as is the case with a wrist watch, a lamp or the earth globe. Also the letter "O" is included in the concept. Therefore, as insignificant those parts may be -dots and dashes- represent signifying signals of a wide evocative spectrum, whether concrete or abstract.

In conclusion, each part of totality is evocative of something different from its overall meaning. Minimal parts of objects, perceived or thought images and words allow us to recall semantic, episodic and autobiographical events of any time in the past, and can even lead us to inspire artistic works and masterpieces, literary or musical works. It is therefore not surprising that a well-known artistic movement is inspired by geometric abstraction, in which every part represents an autonomous signifier from the rest. All sub-signals, as a result are evocative, associative and signifying at the same time.

In language, syllables and evocative letters from other events are decoded from any of its forms: written, spoken or thought. These signals directly penetrate the unconscious as parallel signifiers after having signified at conscious level. They are essential to abstract, generalize, differentiate or symbolize concrete topics. It is a singular activity that takes place in the hippocampus, the areas of the cerebral cortex with linguistic functions (Broca and Wernicke areas) and those responsible for general memory (medial temporal lobe). It all has active interference. The hippocampus main role would be to relay to the cortex whatever has to be recalled, for which its pathways with the linguistic areas of the brain would be the MTL (medial temporal lobe), which hosts the memory. For example, the syllable "mo" is fundamental to assemble many words, among others "mother", "motion" or "mouth", depending on the current circumstance. It can also generate other ideas related to the words "AMplifier", "MOmentum" and so forth, without our brain having to care much about the order of the letters as they appear. The brain can read OM instead of MO, as required by the circumstances. This is how we learned to speak, write, read and think. A syllable is more than a phoneme, it is a signifier in its own right.

CONCLUSION

How we recall is a great enigma, which still has to be unveiled. As I explained before stimuli and sub-stimuli of cortic-hippocampal origin would have an enormous duty in the evocative process starting from selective processes of inclusion into consciousness. The more seemingly "insignificant" a stimulus the higher would be its evocative spectrum. Recalling involves many things and depends on an infinite amount of variables: physical, neurobiological and semantic. In neurocognitive terms, the unconscious operates on the basis of evocative and signifying gradients. Information can saturate the brain and cause stress, and some memories can get masked in the process, making evocation a difficult task. Whether we have higher or lower evocative possibilities will depend on several factors: the degree of neuronal interconnectivity, which neurobiologically speaking, is a determining factor of lower evocative speed. Another factor is age of course; the more advanced it is, the more it would slow down the process, for example the evocation of proper nouns, even if the event to be recalled remains in consciousness. Furthermore, using the same stimulus to signify differently, or visualizing parts of the face of a person, whose name we have forgotten, or places visited by this person will allow to recall him or her promptly under normal conditions, besides signifying something else.

Nevertheless, it is a fact that remembrances weaken over time, making it a difficult task to recall them at conscious level, but in many cases it would be a rather apparent than real shortfall. The brain is submerged in a sea of information in the process of semantic, cognitive and emotional decoding and recoding. The older the age, the more associative capacity is accompanied by higher consciousness of our ego and surroundings, but the lesser the speed at recalling one-off occurrences and specific facts. Be as it may, our brain hosts plenty circuits with the capacity to "accommodate" memories that we regarded as forgotten, but which are actually "masked", so that it will take some time to get hold of them. In the brain the unconscious can be localized at biological level without signifying properties.

Finally the evocative process allows us to retrieve allegedly lost remembrances. Each part of a conscious meaning generates unconscious activity, but not necessarily as a purpose or as a complementary component to itself. Memory and evocation are closely related activities, but they are not the same. Evocation encompasses far more than memory. In other words, at conscious level we can retrieve more than what has been memorized, partly because beyond remembering we also forge subjective meanings.

To summarize, forgetfulness is perhaps one of the most controversial cognitive activities in neuroscience, psychology and psychoanalysis, and its concept is essential and requires revision in order to unveil so many unknowns inherent to this particular field. To forget has the equivalent meaning of the relative incapacity to retrieve a particular remembrance. Notwithstanding this "apparent loss", the brain would not lose information that would induce its memorization, but instead it would keep the forgotten information linked to another information with the aim at preserving it at unconscious level and then restoring it to consciousness for other cognitive, emotional or evocative circumstances. What is true, though, is that short-term memories without consolidation would be lost. What has not yet been proven, however is that identical functional information to other possible meanings –maybe all of them– even if they appeared to have been lost at conscious level, would not be lost in the end.

This means that the unconscious would keep it latent and could even rebuild itself ! And this is the peculiar phenomenon that would pave the way to lay out the necessary groundwork in order to rethink the subject of forgetfulness. As a first summary, we can say that the memory's most prone part to be forgotten would be some specific (selective) data from the remembrance and not precisely its meanings. From this follows that the information contained in the brain would be in a constant redistribution process, depending on the meaning to be recalled in each circumstance.

As a second summary, we describe what forgetting is all about: to forget would be like a sort of "gimmick" of consciousness to preserve memorized information, in order to reuse it for various purposes, however not necessarily for evocative purposes of that memory. Alternatively it would be "used" to rebuild the remembrance that resists recalling. Hence, forgetting is not about a process aiming at freeing neuronal space in order to form new memories, but a way to preserve the neuronal space and thereby contribute towards the long-term consolidation of those memories.

It is natural, for many reasons, that certain memories are more slippery than others, for example those related to historical dates, whose forgetfulness does not involve losing track of the overall context of facts in the past, nor losing the cognition of the national heroes who took part in those feats. From this follows that forgetfulness is related to accidentally remembering something, rather than the essence of the remembrance. Another example, which is often quoted by various authors tells us about forgetting a telephone number once it has been used, thus illustrating the concept of RAM (Random Access Memory). Forgetting something more often carries now the fashionable term of Digital Amnesia.

This does not suggest losing consciousness or the reason why we are using the number, and also why we made the telephone call in the first place. After all, our brain does not memorize all information but preserves it for other meanings. Moreover, it does not even detect it at conscious level, and here we are talking about the phenomena of apperception and non-conscious memory, as we shall see further on.

Hence, to forget is not about the total loss of a remembrance, since each neuronal circuit operates with an infinite amount of meanings. Thereby it is possible that opposing meanings depend on identical information processed within the same cerebral circuit, but distributed and sorted differently in the neuronal network.

This could also be due to the fact that part of the information is busy, so to speak with something else, like recalling a remembrance, an idea or a subjective meaning. Under normal conditions there is no absolute forgetfulness, however some "forged" forgetfulness does exist, probably because it could have been occluded in the process. In other words, identical signifying/sensorial information, which intervened functionally and simultaneously spawning two remembrances with a different meaning, making it temporarily difficult to recall one of them. Alternatively, it could also happen that those parts of totality, which are engaged in a particular remembrance, could be active in the process of devising subjective ideas. To put it the other way round, parts of said information would rather follow a different neuronal pathway as the remembrance to be recalled. Since nothing is redundant in the brain, occasional amnesia can, by all means, be compensated with another function, such as a subjective construction for instance.

While for neuroscience forgetfulness is an objective mechanism, to a greater or lesser extent depending on neuronal connectivity, for psychoanalysis it contains subjective and defensive connotations of the ego such as negation or repression.

To conclude, Ylya Prigogine, a Russian physical chemist and Nobel laureate in Chemistry used to say: "If we divide knowledge, on the one hand the scientific and the humanities on the other, we will never have a human science or humanity that will ever grow with the aid of scientific discoveries".

PART THREE: The Unconscious Mind and Language

The Unconscious Mind

. Concept

. Unconscious, memory and non-conscious memories

. Unconscious and subjective gradients

Language

. Language, memory and non-conscious memory

. Linguistic stimuli

. Language and Evocation

. Linguistic convergence and divergence

PART THREE: The Unconscious Mind and Language

THE UNCONSCIOUS MIND

CONCEPT

The concepts related to the unconscious and subjectivity varies according to the field of study or discipline. Far beyond their etymology both concepts have become the target of study of the philosophy of the mind for many years, but also psychology, psychoanalysis and in recent years neuroscience, which turned out to be its largest enigma. Once these enigmas are unveiled, it would be possible that some failed predictions could possibly change, for example the belief that AI (artificial intelligence) as well as robotics could possibly represent an evolutionary sign of the human brain, which is absurd.

What still remains to be understood about the brain is enormous; among other unknowns we do not seem to understand how the brain produces conscious subjective meanings based on unconscious activity, which of course are unique and vary from person to person. In other words, trying to unveil the nature of human subjectivity, the creation of ideas and abstraction, just to name a few of so many enigmas about the brain and the human cognitive and emotional lives. As we can see this task would eventually involve the introduction of new variables toward deeper research. With this goal in mind a substantial change of paradigm would eventually take over and above all, a new sustainable theory of the human mind and consciousness is of utmost importance at this stage.

From a neuroscientific standpoint both unconscious and subjectivity still carry many unknowns, and their clarification entails the implementation of two variables into this field of research, namely one independent to the phenomenon, which should be capable of determining the semantic content of stimuli to be processed by the brain, and another dependent on the phenomenon as such, which is the cerebral substrate of the meaning. Consciousness as we know it would not be possible without the brain; the same applies to meanings dependent on physical intensity thresholds, nor times of exposure or latency. Neurochemical changes at neuronal level would not be possible either, since these would be common to any normal brain. What makes it more complex is that the only variability would be the one inherent to the independent parameter. This means that a given stimulus would operate as a different signifier on each individual, circumstance and time, i.e. in the signifying action autonomously from the neural action of signifying, in any case this would go beyond signifying diversely.

Paradoxically, subliminal stimuli lacking the necessary threshold to generate brain activity could however, trigger the actual meaning to be sustained by consciousness. How does this happen?

Consciousness as phenomenon depends on two rather opposing factors. On the one hand we have threshold-bearing stimuli with active influence in attentiveness and perception and threshold-independent signifiers on the other. These two would ultimately be directly accountable for the meanings to reach consciousness. This happens in such a way that, while the brain signifying action turns out to be an objective phenomenon common to every normal individual, its emerging meanings would subjectively differ in each case.

Under normal conditions the brain has its own thresholds. In other words, it would vary and adjust them spontaneously according to each adapting need and circumstance. The stimulus threshold is a measurement unit, which was introduced since the early days of experimental physiology and psychology and it was at the time an essential requirement for brain functioning, but it does not render any help in that it does not tell us HOW or WHAT FOR we do signify at conscious level in the first place. Or what is the purpose for brain activity to become a human mind. The wrong approach is to assume that the stimulus' physical nature as well as the resulting neurochemical processes to its action would be determining factors in the variety of meanings for each individual in response to the same objective stimulus.

The knowledge about the brain as a substrate of mind and consciousness should be reviewed and redefined on the grounds that as has been observed at empirical level, the stimulus' thresholds could in many cases be inversely proportional to its signifying possibilities. Hence, the more general and less liminar an objective stimulus (information) is, the higher its unconscious signifying power. So the unconscious signifier would eventually capture those stimuli regardless if they have or do not have the necessary intensity or time of exposure, so that ongoing meanings in progress would get more chances to gain access to consciousness. So, if we insert the signifying nature of the responsible stimulus as a research variable after processes of analysis, synthesis, convergence and divergence, one could reach a point to show that minimal parts of totality without threshold of their own and neglecting attentiveness and conscious perception could eventually trigger the subjective meaning.

Let's then summarize: consciousness can be two things, namely mechanisms that take place starting from threshold-bearing stimuli and meanings, which depend on the signifying nature of the former, as totality and/or parts in any of its degrees or rather depend on the attained brain activity level, which we call gradient. This gradient was reached by undergoing unconscious meanings, which would only require subliminal information to be able to access consciousness. Stimuli consist of multiple parts, which are generally autonomous from totality. Let's think it this way: the latter represent another totality composed of identical parts from the former, but with another time-related and spatial distribution. These parts are out of context from defined totality and would eventually mean something else in subjective terms. In conclusion, thresholds will be necessary to produce brain mechanisms supporting a conscious state, for example the remembrances (memories) but not as an exclusive condition, but to be able to install a subjective meaning instead.

Subjective meanings would rather have unconscious roots and would be generated as minimal parts of the stimulus. Now, how do these achieve signifying action while still being infra-thresholds (subliminal thresholds)? This is precisely the core question in this book. The problem of the mind and the subjective consciousness has always been a "nut hard to crack". Many renowned researchers worldwide sustain: "We do not have the right answers yet" or "It may be possible that we'll never find the answer at all". I personally feel more inclined to think that its solution is utterly feasible.

The attempt of memorizing permanent information or to perform complex calculations within an extremely brief period of time is not our brain's "forte", but it is AI's strength. But this is not a drawback or disadvantage. As I pointed out right at the beginning of this writing, the evolutionary feature as such is the creative capacity of human kind to conceive the technology capable of doing so, and not to calculate or to literally memorize an overwhelming volume of information. Believing the opposite implies confusing the creative process with the created object. That's precisely the reason why I am trying to insist in that whatever still remains unknown about the human brain would allow to redefine the actual projections of AI, which are extremely useful in procedural tasks. However semantic and creative tasks are excluded. Furthermore, the human brain is capable of storing and processing far higher amounts of information, which would require thousands of computers to work, but its downside is that our brain's "working memory" is ephemeral and to signify it uses each and every minimal part at unconscious level.

Albert Einstein was very wise in saying: "Imagination is far more important than knowledge". And he was right. With the imagination, which overwhelmingly abounds in the world of knowledge we would have far less difficulties to overcome, and many of these are more apparent than real. Subjectivity is not governed by standardized patterns of normality or abnormality. Individual semantic diversity is normal in every case and circumstance, even if many meanings are related to questionable values or if these clash with moral and/or ethical principles.

Mario A. Bunge, a bright and distinguished Argentinean epistemologist once said that the Ancient Greeks used to talk a lot about the unconscious mind, which is a true fact. Sigmund Freud was definitely not the first one in realizing this and its relevance. But this could indeed be possible by making use of deep psychology as a more or less active phenomenon than consciousness itself. And this is indeed a phenomenon, which neuro-philosophy is seeking to unveil by means of computer simulations of brain functioning without any foreseeable results (for further reading see: "*Mysteries of Consciousness*" and *"The Mind"* by John Searle et al. and *"Sweet Dreams" by* Daniel Dennett.

Language taken to its minimal expression, such as its elementary building blocks, such as letters, syllables and phonemes or plainly to linguistic symbols in the written form), represents unconscious subjective stimuli of very high evocative and signifying power. And here I would like to bring back Einstein's reflection, which he indeed had put into practice in his theories, ahead of his time showing those in empirical demonstrations. In scientific terms the unconscious should be thought of with more imagination by means of theories far beyond experimentation, but without omitting or excluding it, especially by adapting current methods and techniques to the research of new cognitive needs. That is why it looks promising to consolidate all knowledge about the brain with other fields of knowledge, whose contents are non-related to neuroscience, such as Economics, Decision-making in Politics and even Art and Technology among others. I would like to point out that it is promising if and only if we do not try to oversimplify the concept, for instance, if we make the wrong assumption that better or worse decision-making is due to a higher or lower secretion levels of serotonin and/or other neurotransmitters, which in turn undermine the signifying capacity of the unconscious.

We should therefore review the concepts of the unconscious vs. non-conscious behaviour. And telling apart their differences is of utmost importance; the latter is only a mechanism while the former is just an active signifying system, i.e. a producer of meanings, whether these have conscious status or not.

I would like to quote a passage by the Argentinean journalist and writer Inés Garland (born 1960): "We write with the unconscious and in the process we open up scenes, about which we were unaware we had them stored in our minds. Some of them are autobiographical, others have been told or seen, but they are exactly what I need for this I am telling".

THE UNCONSCIOUS, MEMORY AND NON-CONSCIOUS MEMORY

The unconscious represents a cerebral and cognitive state, which has been studied by several disciplines for a long time. Among those was psychoanalysis, for which the unconscious has played a central role in its many diverse doctrines. Hence, a possible way of unveiling major unknowns of the unconscious and subjectivity would be by establishing correlations of the unconscious with memory and non-conscious memory. For this purpose the focus should be on the semantic nature of the stimulus, but far beyond its physical and chemical features. Stimuli are material entities, which can be either concrete or abstract. They are carriers of objective and subjective meanings, the latter as a product of their endless constituent parts. Let's take for instance any word whose spelling characters in separate form are inducers of infinity of word entries and concepts, in which the source word would not make any sense at all. In other words, they would indeed have a diverse signifying capability of that particular word as a whole, uniquely and unrepeated on each individual.

Stimuli have basically two fundamental functions, which are essential for a better understanding of the cerebral function at both conscious and unconscious levels: first, to generate meanings from memory and second, as constituent parts of non-conscious memory (non-conscious parts of memory), to recall those memories or assigning meanings in a different way from remembrance. In the first case this can be accomplished by means of threshold-bearing stimuli and in the second case meanings can also be generated even if the stimuli are subliminal. In a nutshell, there are two planes of cerebral activity: the conscious one, whose activity will largely depend on stimuli lying between an upper and lower limit of threshold intensity, and an unconscious plane operating by means of subliminal stimuli.

As a result a subjective meaning can be generated from minute parts of information that can have objective meanings from consciousness following the stimulus as an indivisible totality when its meaning lacks interest for individual subjectivity. From this follows that the subjective experience accompanying such condition over time would have been established only by one part of the stimulus, even if it is subliminal.

There is a type of memories called declarative or explicit memories (about which there is a wide range of literature, which I recommend to read) whose main function involves retaining meanings and events of our lives in consciousness. They are the so-called semantic and episodic (autobiographical) memories, which *after having signified* and memorized would become available functionally to re-signify something else than memories. This is perhaps the reason why memory is composed by meanings, which are relative to inducing information as totality and to meanings of fragments of non-conscious memories, which have little or nothing to do with remembrances. Memory, as we will see in the next section is closely associated with subjective signifiers, which were never part of any memory at all. This is indeed a point of intersection and turning point between conscious and unconscious activities. Conscious memories are an inexhaustible source of parts, which are capable of signifying differently from remembrances, so there would be a functional similarity among them with a subjective experience. The same information can release endless meanings, even opposing ones. In other words, homologous parts replicated in the inducing information of these memories would simultaneously detect each infinitesimal part of the stimulus, as non-conscious memory. This might perhaps be the reason why totally unrelated or antagonistic emotions are able to coexist in every individual experience, such as hatred, love, pleasure, suffering, happiness, sorrow, among others.

Brain information would be undergoing a relentless fragmentation process in signifiers of something different. This activity would be tending to cognitive enhancement, since common *parts –after* having signified– would be put at disposal to be able to signify again under other times and circumstances and not necessarily with the same meaning. This is a sort of de-consolidation process of brain information, which –as the name suggests– does not imply the opposite of consolidating. Parts of non-conscious memories would be common (as information) to any memory and can be selectively regrouped into sub-percepts and subcognitions, which would establish a cerebral link between conscious memory and subjectivity. That is, between the stimulus' objective meaning and a subjective experience. It is a mistake to assume that when perceiving we only detect what is relevant to the objective stimulus to be memorized. In actual fact, our sensory system is unable to detect each infinitesimal part of totality only at conscious level, for which the unconscious would be detecting it to signify on its own and would have to struggle so that it reaches consciousness.

Summarizing, our brain does not discard any bit of information even if it is irrelevant to retain an objective meaning as conscious memory. Attentiveness and conscious perception are limited. The unconscious detects everything including minimal details, which in physics and neuroscience are regarded as subliminal. But the controversy between neuroscience and psychoanalysis has come a long way. Fortunately there is a change of trend taking place in this regard. I personally think that by unveiling its signifying stimuli and the resulting cerebral processes being unleashed, both disciplines could work well together. This can be achieved by focusing into the information, which does not get properly decoded by means of conscious attentiveness and perception. This means that by identifying whatever is "imperceptible" with subliminal thresholds, we would be able to show that memory –under natural life conditions– would no longer require ongoing repetitions as in experiments because the subliminal parts contained in each percept or cognitive action, replicated systematically in every stimulus would consolidate that particular memory allowing its long-lasting effect.

In short, all decoded parts of memorized or perceived information represent stimuli, which despite being subliminal bear high signifying power. This is singular phenomenon that is not governed by the reciprocal laws of excitation in biology but by paradoxical phases or states, in which weak stimuli are capable of triggering intense responses and vice versa.

In this section we have to clarify the concept related to explicit or declarative memories. There is a widespread belief in referring to these as being only conscious, which is partly true, since this would only apply to their meanings as a product of totality, i.e. of the summary that each individual makes from each meaning to be memorized. This is, however, not the case with their infinitesimal parts, whose nature is unconscious. They will potentially signify something different from the overall remembrance. Another possibility is that their meanings whether real or potential, may contribute to the genesis of false memories, while associating with pre-existing memories they had originally assisted to generate. Another aspect to be revised about memory is concerning the other sort of memories, namely the implicit or procedural memories (also called semantic and episodic), which have different cerebral substrates than the explicit ones: here again there is a common belief that they are unconscious in nature. This again is partly true, since it only applies to its mechanisms and not for their action and effect, such as walking or any procedural conscious action.

The unconscious goes far beyond mechanisms, a reason why it is worth mentioning that many specialized books and journals or even for public outreach declare through examples that describing whatever has been memorized at conscious level in the course of existence would take a short time, maybe a couple of hours. This is not necessarily true because from a same starting narrative new evocative signifiers and meanings will eventually emerge in cascade. Signifiers and meanings, which we were unaware of having ever thought, perceived or memorized. I think that a lifetime would not be enough for us to recall all our remembrances; those are too many in there as we could possibly think. The unconscious mind does not forget anything; it retains everything.

Our brain is the largest information reservoir in nature and culture. The goal of unconscious information is to feed consciousness with subjective signifiers. We have to bear in mind that information as such is finite but its possible meanings are endless. Therefore, memory is an extremely wealthy source of signifiers that are non-dependent on thresholds, intensity or specific neurochemical changes, such as stimuli. For better understanding of this mechanism there will always be as in all neurobiological process, underlying chemical activities bearing signifiers and subjective meanings, but these would not vary among various individuals. In other words, diverse meanings would have the same neurochemical basis.

Another important aspect to be considered regarding the relationship between memory and the unconscious is related to memory consolidation. As I explained in the previous paragraph memory would not depend –under normal life conditions– of repetitive threshold-bearing stimuli but on minimal stimuli, which could even be subliminal. In other words, memory consolidation would depend on unconscious signifiers implicit in that memory or in others to be consolidated. Therefore the less seemingly relevant a stimulus, which has actively taken part in building up of a large amount of memories, the higher would be its consolidation power as well as its signifying capacity. "Nothing is lost, everything is transformed", as stated in the well-known definition in Physics' laws of conservation of energy. And this also applies to the brain.

The hippocampus plays a fundamental role in memory and would be substantially influential in building up of evocative gradients or signifiers of events, which were never part of a memory. Subjectivity in its pure form has its roots in dreams (oniric episodes in non-REM and REM phases (rapid eye movement) of the dream and at the core of abstract thinking during phases of wakening. In both activities we are free from perceptual restrictions due to over-stimulation or internal impairments.

UNCONSCIOUS AND SUBJECTIVE GRADIENTS

The gradient is a concept from physics, which in our field of study is related to the existing brain activity in the creation of meanings. In other words, in order to ascertain, whether a meaning is able or not to access consciousness. These gradients will increase or decrease their level depending on the stimuli detected by the brain, especially in their infinitesimal parts. This means that all information carrying objective meanings, while being signifying parts of something else, would definitely impact on functional gradients, which underlie unconscious meanings being built at conscious level, although they might never reach this state. Parts of these stimuli while being replicated in all information that has already signified or is undergoing this process, will impact on the level of each gradient. When the brain detects information it starts rigorously classifying bits and pieces, some of them for their eventual memorization of their meaning, while others as subpercepts, subcognitions or underlying thoughts, in order to mitigate between the objective meaning of the stimulus as a whole and the possible signifying action of each part. This is a fundamental activity for the time-related continuity of consciousness.

Considering the phenomenon from this perspective, the brain functioning of the unconscious could be explained on the basis of dynamic gradients which will be either increasing or decreasing depending on objective priorities of subjective individual needs. Stimuli, which could originate from simple colours and shapes, even linguistic symbols (letters), for example would be essential in the conscious conversion of unconscious meanings, even when lacking threshold. This can happen as subliminal signifiers carrying meanings capable of inverting their thresholds; i.e. constituent parts of totality derived from information with objective meaning would make use of their thresholds, or even of brain information generated by that totality by impacting on the generation of subjective meanings.

Let's look into an example related to doubts induced by conscious meanings of stimuli, as a result of whatever has been thought or perceived: To doubt about what we see or think is fundamental for the reasoning of a particular experience. Its solution would sometimes depend on signifying parts, which are implicit in the problem that caused the doubt, since every stimulus is an inextinguishable source of parts capable of signifying something else.   These   are mainly signifiers representing stimuli, which not always carry thresholds and are subliminal. They can be either verbal or non verbal. As I mentioned before, these are implicit or inherent to the fact that generated the actual doubt, although they can also be inserted in circumstantial events coinciding with time. Minimal fragments of what has been read, seen or heard, or even of dreamed phases, while being detected by the brain cell functioning sustaining the problem about which we are questioning something, would eventually assist to clarify a doubt. A simple syllable appearing almost unnoticed out of any seen word, such as "quo" for "quotient", while it has nothing to do with the specific doubt, could induce even without threshold, the word "quotation" having only the common syllable "quo", which is fundamental for us to formulate the question on how to clarify the doubt. This example is seemingly paradoxical but possible. Everything is credible in subjective terms. Surprising is the fact that one could interpret the meaning without paying much attention to the word "quotient", from which we only perceive the syllable "quo".

In a nutshell, subpercepts and subcognitions would eventually contain elements, which will allow us to initiate and develop any subjective experience. In other words, minimal signals broken down from phrases, sentences or ideas, detected by their homologous and replicated in other concepts would eventually become signifiers in their own right. We have for instance the subpercepts "NWP", "HGK" and "MRB" for "Newport", "Hong Kong" and "Marbella" respectively, inducing city names. It can also take place by diverging in order to assess something as "Much Better" or "MaRy", a proper name.

Abstraction represents in neurobiological terms an evolutionary sign, which enables us to think and assign symbols to something concrete without the need of its direct presence.

If we look at the problem from this perspective, the human brain does not waste information in the inductive thinking, regardless of its nature or meaning. What's more it neither does respect the sequence of spoken or written formal characters of the information. It merely uses whatever it needs depending on subjective priorities even it its concealed or "masked" by totality or totally absent from perception. As a result, the order of the constituent parts would be immaterial since the brain has already made use of them as non-conscious memories. Perhaps it could be the actual objective pattern to signify at subjective level. When we dream of a particular person, he or she may resemble someone else, or even an imaginary being "fitted" with parts of several individuals. And it is precisely these parts the ones in charge of signifying at both unconscious and subjective levels or even to consolidate memories and learning processes.

Furthermore, it would no longer be necessary to forget in order to make room for new remembrances. There is the common belief that the brain only works with whatever has been absorbed at conscious level. But it is more complicated than that: all information, which gets memorized at conscious level, is subliminal in relation to what has been perceived. Forgetting would imply that all information responsible information related to remembrances would be busy with another memory or actively involved in the middle of another signifying process. Everything that has been perceived will remain stored in the brain and each part will trigger thoughts and subjective meanings. It is indeed quite a topic in our times, in which an educational reform of our schooling system is looming. As a matter of fact it is imperative and necessary. Wouldn't it be better perhaps to teach how to signify instead of explaining ready-made concepts? Or perhaps furnishing these concepts with parts to build a new totality?

Images would allow less analysis and evocation possibilities than language does, perhaps because their constituent parts hide more easily than those within totality. Unlike images, every word, syllable or letter is neither captive from any linguistic set nor from any specific meaning. And this allows it to insert itself into new concepts and ideas. In short, images and language are essential to signify and each and every part will impact on the gradients of the unconscious. In fact language displays a wider signifying range than images.

In conclusion, the underlying brain activity in both mind and consciousness would rather depend on the stimulus' semantic nature, which is an entity capable of influencing the stimulus' functional gradients or a brain activity generated during the meaning's retention as per adapting subjective needs. Gradients generally define more or less higher or lower possibilities, so that a meaning could acquire conscious state while being governed by the stimulus' features. This is in fact a phenomenon driven by the stimulus' qualities rather than by its threshold intensity. This is a very useful concept to understand how the brain generates a meaning at conscious level. Stimuli, whether they are perceived from the surroundings or built up inside the brain are bits of information that replicate in a different functional order within any possible stimulus. In other words, any stimulus independent from its meaning if taken to various breakdown levels bears identical parts the higher its resolution grade. Hence, every time our brain is stimulated each part would be detected by its homologous one, which in turn is replicated in other meanings (both conscious and unconscious) while impacting on their respective gradients. In terms of consciousness, totality is not the sum of all constituent parts. That is the cause of its enormous influences in the consolidation and evocation of memory and learning, as well its importance in building up subjective meanings.

LANGUAGE

In all its forms, language whether spoken, thought, read, written, seen or touched (as a communication means for the blind), with gestures or movements (for the deaf and dumb) is definitely an exclusively evolutionary sign of the human kind. It represents an essential cognitive instrument in human interaction and communication. Among many other features, its function entails abstracting the meaning of concrete objects to be thought about, without the need of their direct presence. The language's main enigma is it intricacy related to the cerebral nature of the process of sparking ideas prior to linguistic elaboration. In other words, how ideas become words. Language is associated to images and these in turn while inducing ideas, would be their stimuli. Language is fundamental in communication in self-knowledge, our ego's consciousness (self-consciousness) and the world surrounding us (hetero-consciousness).

The brain unlike all other organs of the body is not governed by standardized patterns of normality of abnormality. The heart, for instance for the sake of its normal functioning, depends on innumerable factors such as the nervous system driven regulation contractibility, its pace, frequency or the pressure exerted by the blood onto the blood vessels at each systolic or diastolic cycle. The brain, on the contrary as a system in charge of signifying at subjective level is not exclusively driven by strictly physical factors, such as the threshold of stimuli or by chemical processes (different ions or proteins for every meaning). But it is in fact governed by sociocultural and ethical factors. Furthermore, identical neurochemical processes would rather support the variety of infinite concepts intervening in language. That is the reason for encountering empirical difficulties for determining its essence and true nature. Different meanings in response to the same stimulus are no equivalent of normality standard or pathology. With language we have the same situation, since a given word can vary its meaning according to the overall verbal context. All information or stimuli has a decisive influence in language. This is how syllables, phonemes, linguistic symbols or letters, images and sub-images play a central role and determine the concept and verbal production. Every brain can signify differently in response to the same given stimulus. The word and its parts besides bearing one or more meanings, can operate as a signifier. Summarizing, language is the main functional pattern of subjective consciousness. Therefore, language and images are determining factors for the uniqueness of subjective meanings. In this regard, in the opinion of the Czech film director Harum Farocki certain images are distorted and even immoral. In the end I think that this can also happen with the spoken language. Farocki wrote a book entitled *"Distrusting the image" in which he is* utterly critical regarding certain type of images. I quote a passage: "We have to be less naive when we see something". To my understanding I insist that in terms of language the difference is that in spoken language, parts of totality, such as symbols and signs would have more possibilities of falling out of context, due to their higher degree of signifying volatility, than the parts of an image generally more merged into the meaning of the central figure, as is the case with a syllable or a word as part of a text. Therefore its meanings would not remain captive without having the possibility of expressing themselves in consciousness. With the image it would be the other way round, since totality disguises or "masks" its parts. Words, syllables and letters on the other hand would have a higher cognitive and signifying fluidity. They are not tied up or attached, so to speak to any meaning. This is perhaps the reason why at unconscious level, verbal parts would signify without major difficulties in relation to the non-verbal ones. Abstract art would be the exception, maybe because every part of the work, apart from being functional to totality would be able to signify autonomously and diversely from it.

In conclusion, every verbal composition and its structural and phonetic parts are common to diverse meanings but not strictly attached to them. Every spelling character, whether words, syllables or letters is a cognitive instrument of abstraction and keeps its autonomy to signify differently from totality. At subjective level the difference would rather be the higher possibilities of the parts of language to diverge into other concepts. Sub-Imagery, on the other hand, would have fewer possibilities of convergence and divergence in other contexts.

Another aspect to be taken into account is that the brain works based on the information, which after being signified gets recycles into fragmentary parts to infinitesimal proportions without being trapped by the generated concept. Therefore, it would allow it to keep on signifying diversely, by converging and diverging into subjective meanings undergoing unconscious elaboration. Wrapping up, words as well as images take part functionally in the creation of ideas, which will happen before and after the linguistic production.

From the moment we were born we started joining and linking images and words. During the first year of life we associate objects, which are linked directly interlinked, like the concrete presence of our mother with the nourishment, such as the breast or the baby bottle. But once we have acquired the skill to designate these verbally (abstraction), the child no longer will make use of the concrete concept by replacing it with words, namely mother or milk. Abstraction would be the mechanism allowing the brain to function in phase, according to the state adopted by the stimulus. In this way it will no longer require intensity thresholds, thus inverting the egalitarian laws of biology of reciprocity of excitation.

LANGUAGE, MEMORY AND NON-CONSCIOUS MEMORY

The unconscious such as the language is related to memory and non-conscious memory. The brain beyond being intensely populated by neurons and activity thresholds, is also fully packed with signifying information at verbal and non-verbal level (unconscious) We think and communicate without having to recall every single word. Jacques Lacan and Carl G. Jung paved the way in that they were able to recognize the importance of the language and thus its singular characteristics in the unconscious realm. Jung's approach was from verbal associations, the thorough study of symbols and archetypes, while Lacan had brought forward the idea that the unconscious operated with a language of its own.

Letters of the alphabet are classified in two groups, namely consonants and vowels, the latter with a higher phonetic resonance and the former,–although not as resonant as vowels– with other thresholds opening greater possibilities of neuronal diffusion for the genesis of ideas and language as a means of expression. Absolute meanings are of no use at all. When consonants and vowels merge they form syllables and phonemes, which will eventually converge or diverge into new words, ideas and concepts. Furthermore, the more elementary these syllables and phonemes are, the higher its linguistic and semantic inclusion power will be, especially in various meanings, therefore its importance in time-related continuity of meanings and consciousness.

Language is an essential tool for signifying and re-signifying. The brain does not discard information, and even less if it is of linguistic nature. Any specific word entry can be built out of the most diverse cognitive and emotional states of mind. Within the context of love "the letter "M" may lead us to the syllable groups "MOM" for "MOther"; within a context of violence it could lead us to "MUrder" and within a playful context to "MOcking", for instance. Any concept, phrase, word, syllable or letter releases subjective stimuli, which are necessary to spark new ideas, even if the resulting meaning has little or no connection with the ongoing brainstorming process at all.

LINGUISTIC STIMULI

Stimuli of diverse origin and nature represent the raw material of brain activity, the functional pattern of subjective consciousness. Stimuli are entities, which can carry multiple meanings, such as objective and subjective meanings. There is an exception though, because the objective meaning of totality will not necessarily be the one accompanying a conscious state. Let's consider for example the following: the result of a sports event may or may not be of our interest, no matter how strong it gets broadcasted through all possible communication media. However it is possible, that we'll never find out who actually the winner or loser was. However, we could learn about a friend who becomes emotionally involved in it (joyful or sad about the outcome). And this will be the conceptual content, which will fill our consciousness.

Stimuli may come from two different sources: exo-cerebral, that is when their origin lies outside the brain and endo-cerebral, when they are released by the brain's own functioning without direct intervention of the senses of attentiveness and perception. Its nature, on the other hand is utterly varied and encompasses the totality of integration levels of matter taking place in biology, in physical chemistry, in psychology and last but not least in the socially cultural environment. Semantic diversity is the parameter for a better tentative understanding the reason why subjective meanings come into being in the first place.

Exocerebral stimuli

These are stimuli being generated outside the brain, either in the environment or inside the body (organs, muscles, and so forth) but not inside the brain. As I pointed out before, images –whether abstract or figurative– are the most complex exocerebral stimuli and the words in their most diverse forms: written, spoken read, heard or "performed", and so forth. The brain produces language based on information detected by our senses, which after being transmitted as nervous pulses to centres of higher complexity will remain encoded within its neuronal networks. *At a later stage it will* eventually be decoded and re-encoded into parts whose goal is, among others, to generate subjective experiences. As a result, all stimuli once decoded in the brain will break down into smaller components, which will be sorted into new stimuli, but not always with a threshold, i.e. some of them at conscious and others at unconscious level. Their aim is to signify autonomously the perceived totality or the memory as such. Stimuli generated in the body, among other structures all viscera and muscles also fall under this category and they are known as visceral receptors and muscular receptors respectively and are indirectly linked to language. Let's look at the following example: hunger or thirst induce us to speak up the sentence "I am hungry" or "I am thirsty" in order to convey the message across. Furthermore, in the case of stimuli originating from the locomotion system are of great importance in attitudes, postures and impulses, as well as in implicit memories.

Intra-cerebral stimuli

These stimuli are generated within the intricate environment of neuronal circuits and work independently from attentiveness or perception. They represent "silent" signals, which are generally conscious and subliminal, although they are rather connected with brainstorming, comprehension and language production. Their nature is abstract but not immaterial and in many cases they lack thresholds of their own and generate conscious meanings due to the fact that the brain already has its own thresholds and is being continually stimulated. Brain activity is liminar by nature. At any given moment our brain is capable of processing vast torrents of liminar and conscious amounts of information, which are hard to measure. And from there we extract sub-percepts and sub-cognitions, whose aim is to re-sort ideas and meanings. Conscious thoughts are not always carriers of subjective and relevant meanings. This is why we only choose some of its constituent parts out of memory, in order to be able to signify or for verbal communication purposes.

Summarizing, linguistic stimuli converge into words and ideas during the processes of conception of ideas and phonetization. Despite the fact that the mentioned components (words and syllables) might be subliminal or lay subliminized by the verbal set, they will still diverge into new concepts. Every sub-stimulus becomes a re-signifier of the previous one. Stimuli are neither intense nor slight during re-signifying and subjectivizing actions. They are not quantifiable either because their essence is qualitative and unrepeatable for each individual.

Environmental stimuli, like intra-cerebral stimuli, are constantly fragmenting into smaller parts aiming at individual subjective needs.

Linguistic sub-stimuli

The sub-stimulus concept arises from the fragmentation of brain-related information. A linguistic sub-stimulus represents those parts of language such as words, syllables, letters and phonemes, which will be in charge of generating new ideas and meanings in parallel with the meaning of totality at both conscious and unconscious levels. This is accomplished in such a way that ideas and language retain their continuity over time. Minimal particles of information will be in charge of supplying all the necessary elements, in order to deepen or modify them accordingly. In a nutshell, linguistic sub-stimuli may either be originating from perception or from the brain and would be ultimately responsible for the meaning's emotional content.

I would like to quote Serge Haroche, Nobel Prize Laureate in Physics in 2012: "...if we keep on deepening our focus into the states of matter, we are going to end up in parallel universes".

LANGUAGE AND EVOCATION

Language basically consists of words and these in turn are composed by syllables and letters, linguistic symbols and phonemes of a very wide signifying power, independent from the implicit meaning of the words to which they belong. In other words, they hold a very high evocative power of ideas and concepts. The mode of operation would take place by means of inclusion of new word entries to develop the idea in progress. For example, the syllables "CA" and "PE" are directly evocative "CAR", "CAPture" or "CApE", among others. As we can see, minimal variations in the sequence of linguistic characters or symbols cause the meaning of a concept to change substantially. The recalling or evocation process is functional to various cognitive activities such as memory, thoughts and language, among others. Hence, it is also functional to the subjective meanings held by consciousness. As I mentioned at the beginning, parts of a language and its meanings represent an inextinguishable source of signifying stimuli.

In short, words and meanings can be recalled from their own linguistic components. Any word, written in uppercase or lowercase is a sign of abstraction, since we use it to designate or label any object, even if it lies outside our field of perception parallel to other events, and can displace the objective meaning that has been thought of, out of consciousness. This is a paradoxical fact, as the word may lack semantic relationship with the recalled fact or the meaning thereof.

At the same time every letter has a particular shape and contours and could induce concrete facts. For example the capital letter "A" could be associated with the facade of an alpine house, and as an abstraction symbol it could also generate a specific thought. Every letter of the alphabet apart from being a component of totality, which could be a word or a linguistic essay representing ideas, is also a totality in itself. And each and every individual for the sake of signifying would also do it based on that part, if and only if it corresponds with the relevant subjective priorities. As a result, minute and concealed parts could eventually signify differently from him. Subjectivity is therefore the result of a succession of signifying sylogisms, among other inducing signals of minimal parts, such as a written letter still without threshold, for instance. Consciousness therefore should not have any semantic loopholes, since it will never cease signifying under normal conditions , not even during dream phases.

This peculiarity can lead us to the assumption that subliminal stimuli could allegedly "reinforce" themselves during unconscious brain functioning, which makes perfect sense. We should however not interpret "reinforcement" as an increase in size, but rather as a didactic "trick" to explain how our unconscious would process subliminal captured parts of totality by transferring them to consciousness. And how those parts would leave room for further meanings. For better understanding, let's imagine that if the underlying information within any given meaning being generated would capture the minimum necessary amount of information to signify, whether the stimulus has a threshold or not. Another example but this time as a verbal message: if in a written text we perceive only one word or part thereof, among hundreds, we often realize that it displaces the objective meaning of the narrative and that consciousness likely bears another meaning. How could this happen? The stimulus even without the necessary physical threshold, simply made itself "visible" somehow at conscious level, since it was the carrier of the necessary signifying power. This phenomenon would also take place during the oniric episodes (dreams) as we will see further on. Another possibility to explain this activity in a metaphorical sense, would be by examining the detecting capacity of the unconscious in order to capture minimal parts from totality. If we keep on looking from a metaphorical standpoint, let's consider the wide variety of shapes of the regions of planet Earth as seen from satellites at different altitudes. In another scale range let's also consider the shapes and colours adopted by micro organisms visualized through the microscope. These shapes could even resemble masterpieces by an artist.

In this regard I also would like to quote the Italian writer Antonio Porchia (1886-1968): "What words express does not last. But the words are everlasting. But the words are everlasting, because what they do convey is never the same."

Our consciousness would ultimately signify in a phonological and linguistic manner from minimal parts of totality, while rebuilding itself into new broken down units with parts converging and diverging into new words and concepts. Thus without solution of continuity and remaining unsolved. The fact that we are unaware of the totality of perceived, recalled or thought parts does not necessarily mean that the brain would not "see" or record those parts without the need of an intensity threshold or time of exposure as a stimulus. This process would take place based on the fact that totality plus constituent parts would be detected in parallel, so the brain would only prioritize whatever is essential in order to signify its conversion into the conscious state. For example, the letter "m" to recall "meal" when our basic need is to get something to eat; "my", a possessive pronoun in order to talk about something related to us; or "mattress", when we feel drowsy and want to go to bed. All meanings are possible in a subjective context.

As a result, generation of ideas, brainstorming and language would only originate in processes of analysis and synthesis related to what has been perceived or thought.

The reordering procedure of syllables, letters, shapes and colours, which is carried out by our consciousness is casual and can be the cause at the same time. Each letter is a signifying verbal stimulus, but as I pointed out before it can also be non-verbal. Its power is both evocative and signifying at the same time. For example, the initials "ERQ" composed by one vowel and two consonants can lead us to recall the name "EnRiQue" by convergence of a proper name, or rather "Ed should Remain Quiet" due to linguistic divergence of each symbol in charge of recalling or producing a different word; a so called grapheme.

When a sub-percept joins vowels, its linguistic construction is more direct according to subjective priorities, for example the syllable "DU" could be included without modifications in the name "DUane" or simply "DU". As can be seen, consonants or vowels in pairs would establish direct linguistic constructions, in which the syllables are inserted into the word unaltered, or in the recalled concept. This is the reason for the semantic paradox in various phases or states adopted by the stimulus. Hence, components of sub-percepts or sub-cognitions may branch out into two or more signifying and evocative stimuli, like for instance DU for "Democratic Union" "green meDOws" and many more. I would like to emphasize that evocative possibilities are vast in subjective terms.

Subjectively speaking, our brain would operate by truncating information, that is without using the entire stimulus, but only making use of the minimum necessary to recall, speak up or to signify." It would just use "DU" to build "DUane", since it only would require that part of the stimulus to be able to signify. This is certainly paradoxical, since the more letters contained in a phoneme, the more restricted its evocative spectrum would be. This is because what has been recalled pre-exists in a state of chaos within the stimulus as a whole (all letters). In order to generate ideas at linguistic level, our brain would perfectly do without overwhelming amounts of information, since many parts are stored in itself as non-conscious memories or as conventional memory, so that we only would perceive what is essential.

LINGUISTIC CONVERGENCE AND DIVERGENCE

Thus the brain recalls and builds ideas and subjective concepts by means of processes of convergence and divergence of the linguistic symbols inside the neuronal network. We refer to convergence processes when the words we think about make up linguistic characters thereby building a new word, so that a meaning can get the necessary gradient to become conscious. When we talk about divergence processes, it's because the meaning at *a later stage will* diverge into other sets of words with meaning. As we can see, these processes will enable consciousness to signify without any disruptions by feeding back through convergence and divergence of the linguistic symbols in the brain, consistently to its fragmentarity into parts, so that consciousness can work without interruption. In other words, the aim here is that consciousness should not stop producing or ceasing to bear meanings not even during the sleep phases.

Thus, it is perfectly understandable to assume that language gets generated based on infinitesimal non-conscious parts. The main role of these parts is to restructure the abounding chaos in brain information by converging and diverging, so that objective and subjective meanings can emerge. Wrapping up, constituent parts, syllables, sounds and letters from a grapheme will be able to diverge into words and ideas which have nothing to do and are not complementary to the meaning of the original grapheme. For example, the concept "...a new life..." is part of the necessary signifiers to supplement the idea in progress or to replace it by another one, like thinking that life is good or not worth at all. Or to converge into words with different or opposing meanings. From this follows that the contents of subjective consciousness would depend on convergence and divergence, analysis, synthesis, concentration and scattering of verbal and non-verbal stimuli. While linguistic stimuli would represent words and images, functional processes on the other hand would be of convergence and divergence.

The genesis of subjectivity is of psychological nature and it contains its own logic. Its meanings are unique and not repeated in other individuals. This also happens on a same individual under different times and circumstances. And just by knowing its brain cell mechanisms (whose physical and chemical nature are common to any meaning-inducing stimulus) would not be enough to understand its cerebral basis.

Hence, without attaching the semantic variable into the investigation, i.e. those parts of the stimulus in charge of meaning generation, it won't be possible to ascertain whether those correspond or not with pre-existing meanings in progress of conscious gestation. In physical, chemical and biological terms all stimuli produce analogous activities in different individuals and involve identical brain areas of activity, in spite of the potential semantic diversity of stimuli. For example those areas related to pleasure due to substantially different situations like substance abuse, physical intercourse or messages using mobile phones. In other terms, the activity is still the same but what differs is the actual meaning.

Subjectivity as a result, would only depend on the inverse mechanisms of an objective brain mechanism, since they would only be the product of some of the stimuli. These parts of the stimulus are the ones, which each individual chooses for signifying purposes, while the stimulus as a whole would have the functional task to generate awareness as a biological and cognitive state; or to signify in objective terms.

To summarize, convergence would only take place when letters and syllables detached from ideas and phrases build up new linguistic essays, which will eventually diverge into concepts, which are not necessarily related to, or complementary, or even antagonistic. Therefore, the more univocal the meaning of a grapheme or phoneme, the more restricted will be the probability for signifying autonomously from any infinitesimal fragment of source information. In other words, those signifiers of that information reaching the unconscious would eventually face greater difficulties and interference to signify at conscious level but uniquely to each individual, circumstance or time. In doing so it would take place blocking out all subjectivity, so to speak. A simple meaningless phoneme on the contrary could induce infinite meanings. Wrapping up, meaningless phonemes capable of reaching the unconscious in a direct way would have the possibility of tracking and finding the necessary information to generate subjective concepts and thus develop the meaning as such. The reader may well ask (understandably), whether all signifiers are ethically acceptable or not. This will of course depend on the free will and the moral circumstances and values surrounding each individual.

In synthesis, the actual roots of subjectivity would be unconscious in nature and would be drawn up preliminarily into the brain function as non-conscious memories. For example, any given set of two initials could possibly build a myriad of memories or not memorized concepts, so that when the words formed by that syllable and others could give shape to other words and concepts. As we see, meanings do not correspond in a coherent way among themselves, because they depend on subliminal stimuli or letters immersed in a vast ocean of linguistic information that the less something means, the higher its signifying amplitude would become. In our case of a sequence of two initials, these are capable of generating multiple linguistic constructions independent of their order.

In a nutshell, whatever their origin linguistic characters (spoken, written or thought) are responsible for the sparking of new ideas and conceptual evocation. This works similarly like truncating sections of words, without detriment of the signifying capacity of language. Truncated keywords can be useful to induce or to recall ideas or to think, but can lead to language degradation. For example if we say "congrats" instead of congratulations. The brain does neither detect nor memorize everything –among other reasons– in order to avoid information overflow. It already has plentiful information to work with. Nevertheless, it does not discard anything, which could make use of in the future. It stores everything and does not forget at unconscious level.